Photosensitization of Porphyrins and Phthalocyanines

Photosensitization of Porphyrins and Phthalocyanines

Ichiro Okura
Professor, Tokyo Institute of Technology, Yokohama, Japan

CRC Press
Taylor & Francis Group
Boca Raton London New York

CRC Press is an imprint of the
Taylor & Francis Group, an **informa** business

2000

CRC Press
Taylor & Francis Group
6000 Broken Sound Parkway NW, Suite 300
Boca Raton, FL 33487-2742

First issued in paperback 2019

Copyright © 2000 by Ichiru Okuro
CRC Press is an imprint of Taylor & Francis Group, an Informa business

No claim to original U.S. Government works

ISBN-13: 978-90-5699-354-2 (hbk)
ISBN-13: 978-0-367-39698-5 (pbk)

British Library Cataloguing in Publication Data
A catalogue record for this book is available from the British Library.

Visit the Taylor & Francis Web site at
http://www.taylorandfrancis.com

and the CRC Press Web site at
http://www.crcpress.com

To my wife, Mariko

Contents

Preface

Photochemistry of coordination compounds has been making rapid progress over the past two decades, and tailor-made coordination compounds with the desired properties and multi-functions have now become available. Such advances are most evident with the coordination complexes of porphyrins and phthalocyanines. The chemical properties of porphyrins and phthalocyanines have strongly influenced many fields, especially the following:

1. Biomimetics: Many types of metalloporphyrins and related compounds serve as functional groups of metalloproteins in living cells and play important roles such as electron carrier, oxygen carrier, biocatalyst, and so on. To understand the bioprocess mechanisms, e.g., photosynthesis, some analogues of metalloproteins are synthesized and characterized, and the reaction mechanisms are clarifed by using these analogues.

2. Photosensitizers: As porphyrins and phthalocyanines have absorption bands in the visible region, these compounds have been used as photosensitizers for visible light. Recently, more efficient porphyrins and phthalocyanines have been synthesized for the conversion of solar energy to chemical energy, photodynamic therapy and optical sensors.

As for photoenergy conversion, utilization of photosensitizers is regarded as one of the most promising approaches for creating renewable energy resources from sunlight. Many proposals have been made regarding generating chemical fuels from readily available raw materials, i.e., production of dihydrogen from the photosplitting of water. Artificial photosynthetic systems to produce energy resources are important targets to be realized by utilizing photocatalysts such as porphyrins and phthalocyanines.

Although there is strong demand for a reference book covering the scientific background and recent developments in research on porphyrins and phthalocyanines, there are very few available. The present monograph covers this scientific background as well as applications of these compounds, especially the application for photosensitizers. The monograph also includes reviews of recent advances in fundamental research and application in this field.

The monograph consists of two parts. Part I presents a brief introduction of the principles of porphyrins and phthalocyanines, and the preparation methods and properties of these compounds. Part II represents the application of porphyrins and phthalocyanines as photosensitizers, such as photosensitizers for conversion of solar energy to chemical energy, for photodynamic therapy, and for optical sensors, in which the main figures and tables are based on data obtained in the author's laboratory. The highlights of these exciting advances may help as timely and convenient reference sources.

This monograph will appeal to academic and industrial researchers, graduate and senior undergraduate students in the fields of biomimetic chemistry, photochemistry, and photocatalysts, and researchers in some branches of clinical medicine.

The publication of this monograph is supported in part by a Grant-in-Aid for Publication of Scientific Research Results, Grant-in-Aid for Scientific Research, Japan Society for the Promotion of Science in 1999.

It is a great pleasure to acknowledge Dr. Toshiaki Kamachi, Dr. Yutaka Amao, and Mr. Shun-ichiro Ogura, who have directly or indirectly helped in the writing of this monograph.

Finally, I would like to express my gratitude to Ms. Cecilia M. Hamagami and Mr. Ippei Ohta of Kodansha Scientific Ltd. for their invaluable assistance in the preparation of the English manuscript.

Comments on the presentation from readers are very welcome.

Ichiro Okura

Part I
Synthesis and Characterization of
Porphyrins and Phthalocyanines

Chapter 1   Preparation of Porphyrins and Phthalocyanines

1.1   General Methods for Preparation of Porphyrins

Porphyrins are classified into three categories: (A) porphyrins prepared from natural heme, (B) alkyl-substituted porphyrins and (C) *meso*-tetraphenylporphyrin and its derivatives. In this section, the preparation methods of the representative porphyrins in each category are described. The preparation methods of water-soluble porphyrins (D), metalloporphyrins (E) and chlorin derivatives (F) are also provided.

1.1.1 Synthesis of porphyrins from natural heme (Scheme 1-1)

Protoporphyrin IX (**1**) is prepared from protohemin as a starting material. Iron ion in the protohemin is removed by refluxing with iron in formic acid. After demetallation, protoporphyrin IX is obtained by the addition of ammonium acetate. Protoporphyrin IX methylester (**2**) is synthesized by esterification of protoporphyrin IX in HCl-methanol solution.[1]

Hematoporphyrin IX (**3**) is prepared as follows. Protoporphyrin IX is treated with HBr-acetic acid to convert the vinyl group to a hydroxyl group and is then neutralized by NaOH. By the addition of excess NaOH solution, hematoporphyrin IX is obtained as a precipitate.

meso-Porphyrin IX (**4**) dimethylester is prepared as follows.[2] The vinyl group of hematoporphyrin IX dimethylester is reduced to an ethyl group by hydrogen gas catalyzed by

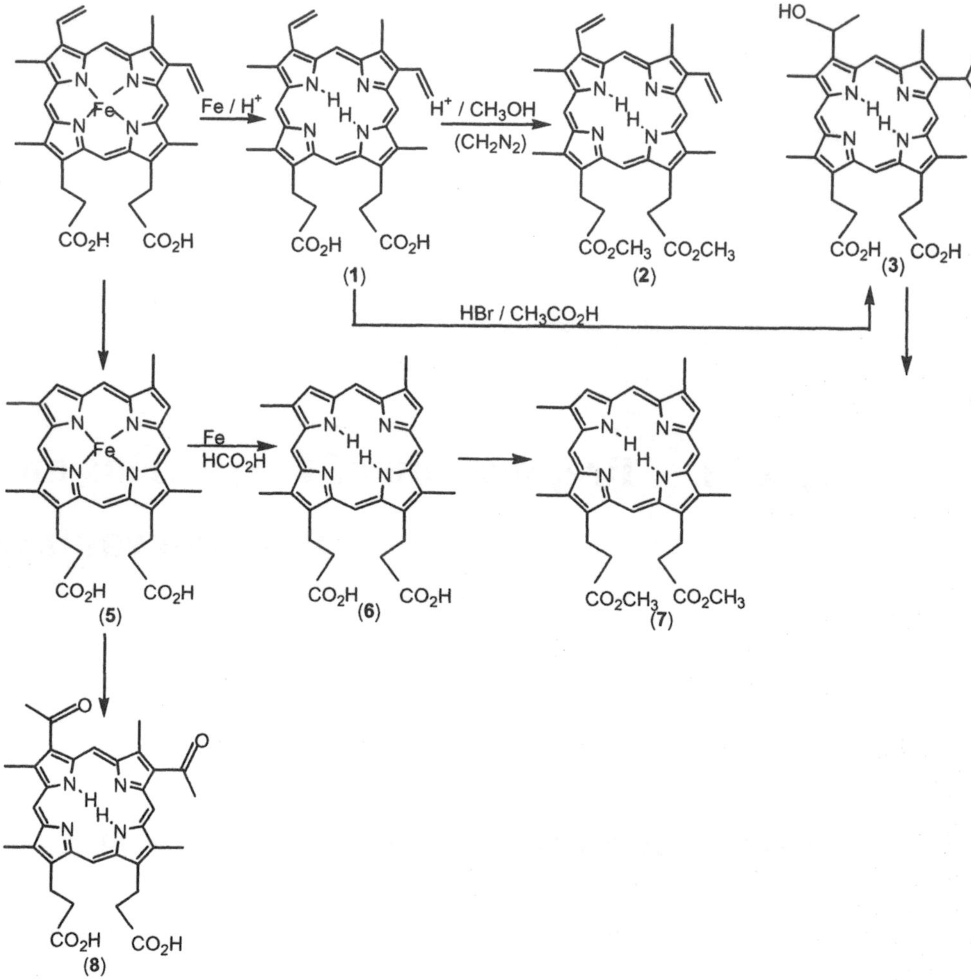

Scheme 1-1 Synthesis of porphyrins from natural heme.

palladium-active carbon under methylmethacryrate and formic acid.   *meso*-Porphyrin IX dimethylester is easily recrystallized from chloroform-methanol. This compound is most popular among the porphyrins from natural heme.

Deuteroporphyrin IX (**5**) is prepared from protohemin as the starting material.[3] Deuterohemin is produced by refluxing protohemin with resorcinol. Iron ion in deuterohemin is removed by refluxing with iron in formic acid. After demetallation, deuteroporphyrin IX methylester is obtained by esterification of protoporphyrin IX in HCl-methanol solution. 2,4-

Diacetyldeuteroporphyrin IX is synthesized by the acylation of deuterohemin treated with acetic anhydride and tin chloride. [3]

2,4-Diformyldeuteroporphyrin IX is prepared by the oxidation of protoporphyrin IX dimethylester using $KMnO_4$/ $MgSO_4$ as oxidizing reagents in acetone. 2,4-Diformyldeuteroporphyrin IX is insoluble in organic solvents. [4]

1.1.2 Synthesis of alkyl-substituted porphyrins

The representative alkyl-substituted porphyrins are etioporphyrin (EtioP) and octaethylporphyrin (OEP). [5] There are three synthesis routes of OEP, as shown in Scheme 1-2.

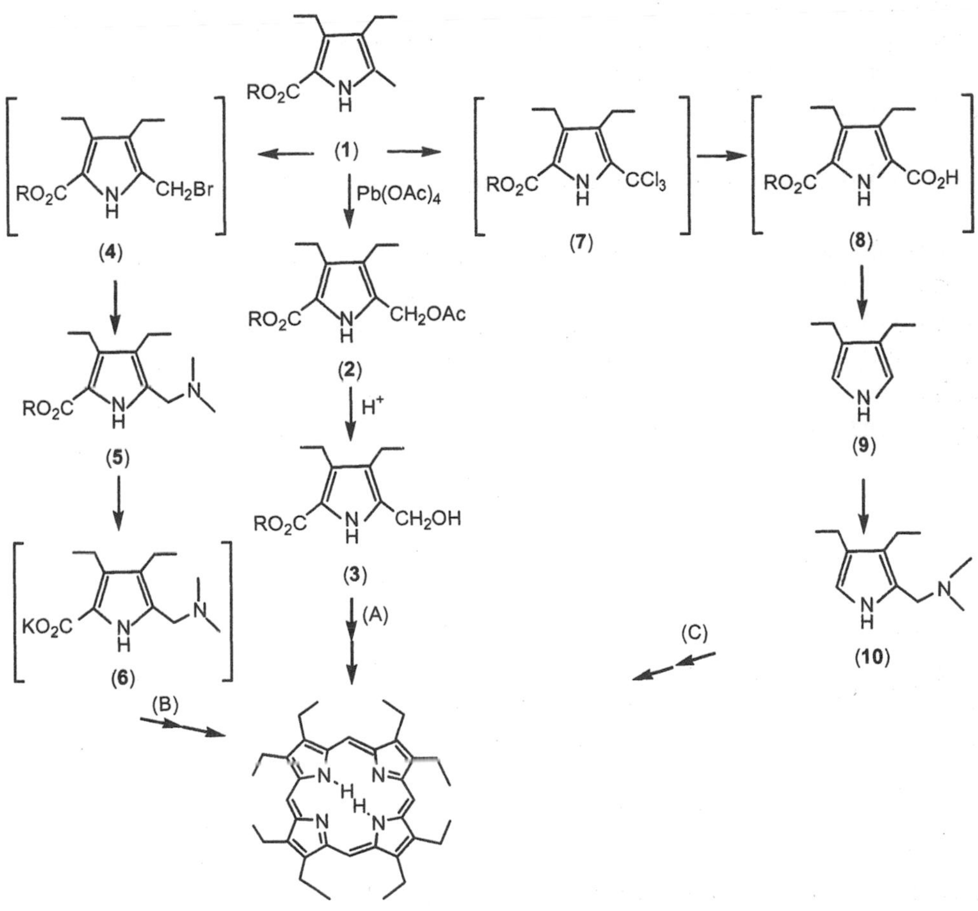

Scheme 1-2 Synthesis routes of OEP.

Route A[5]: The derivative (**2**) is synthesized by the acylation of α-methyl of pyrrole derivative (**1**) using Pb(OAc)$_4$. The derivative (**3**) is prepared by hydrolysis of (**2**) in alkaline solution. OEP is synthesized by cyclization and oxidation of (**3**). The yield is about 20-30%. OEP is purified using alumina column chromatography and recrystallization from chloroform and methanol.

Route B[6]: The derivative (**4**) is synthesized by bromination of (**1**). The diethylamino-methyl derivative (**5**) is synthesized by nucleophilic substitution of amine to (**4**). The potassium salt (**6**) is obtained by hydrolysis of (**5**). OEP is synthesized by cyclization and oxidation of (**6**). The yield is about 20-30%. OEP is purified using alumina column chromatography and recrystal-lization from chloroform and methanol.

Route C[7]: 3,4-Diethylpyrrol (**9**) is prepared by decarboxylation of (**8**). The derivative (**10**) is synthesized by aminomethylation of (**9**) in formaldehyde and dimethylformamide. OEP is prepared by cyclization and oxidation of (**10**). The yield is about 20-30%. OEP is purified using alumina column chromatography and recrystallization from chloroform and methanol.

Octamethylporphyrin (OMP) is synthesized using the above methods.[8] The yield of OMP is very low and OMP is insoluble in organic solvent.

1.1.3 Synthesis of *meso*-tetraphenylporphyrins

Rothemund synthesized *meso*-tetraphenylporphyrin(TPP) by condensation between pyrrole and benzaldehyde.[9] Since this report, *meso*-tetraphenylporphyrin derivatives are synthesized using benzaldehyde derivatives. The yield of porphyrin improves especially when propionic acid is used as a solvent. Porphyrins are usually purified using column chromatography (silica gel or almina).[10] The synthesis route is shown in Scheme 1-3. Tetranitrophenylporphyrin, tetratolylporphyrin, tetra(4-pyridyl) porphyrin and tetra(pentafluorophenyl)porphyrin are synthesized by this method.

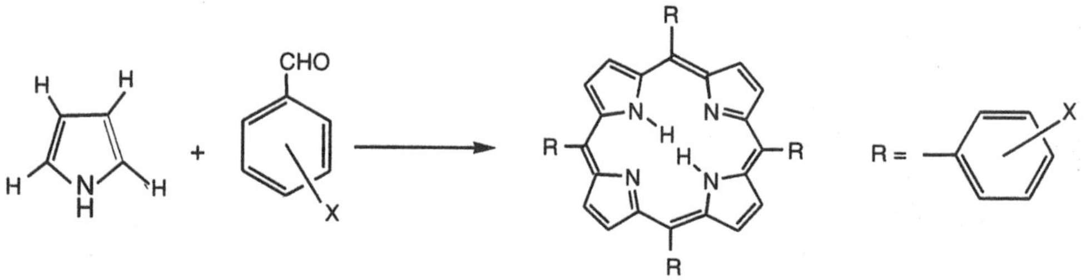

Scheme 1-3 Synthesis routes of *meso*-tetraphenylporphyrin derivatives.

1.1.4 Synthesis of water-soluble porphyrins

Water-soluble porphyrins, tetraphenylporphyrin tetrasulfonate (TPPS), is synthesized by the sulfonation of TPP using sulfuric acid. Tetrakis-(aminophenyl)porphyrin (TAPP) is synthesized by the reduction of tetrakis(4-nitrophenyl) porphyrin under Sn / HCl condition. Tetrakis-(4-carboxyphenyl)porphyrin (TCPP) is synthesized by hydrolysis of tetrakis-(4-methyl-carboxyphenyl)-porphyrin in alkaline alcohol solution. Tetrakis-(4-methylpyridyl)porphyrin (TMPyP) is prepared by quaterization of tetra(4-pyridyl)porphyrin with methyliodide or *p*-toluenesulfonate methylester in DMF.

1.1.5 Synthesis of metalloporphyrins

Metalloporphyrins are widely used as model compounds of metal enzymes and chromophores like chlorophyll in the biochemical field. The standard preparation methods of metalloporphyrins are classified as follows: organic acid method, organic base method and acetylacetonate method (Scheme 1-4).

$$Por\ H_2\ +\ M^{2+}\ \rightleftharpoons\ Por\ M\ +\ 2H^+ \qquad\qquad (a)$$

$$Por\ H_2\ +\ M(acac)_m\ \longrightarrow\ Por\text{-}M(acac)_{m-2}\ +\ 2acac \qquad (b)$$

$$M\,(0)\ +\ Por\ H_2\ \longrightarrow\ [Por\ H\text{-}M^I]\ +\ 1/2\ H_2$$

$$[Por\ H\text{-}M^I]\ \longrightarrow\ Por\ \text{-}M^{II}\ +\ 1/2H_2 \qquad\qquad (c)$$

Scheme 1-4 Standard preparation methods of metalloporphyrins.

(1)   Organic acid method (Zn, Cu, Ni, Co, Fe, Mn, Ag, In, V, Hg, Tl, Sn, Pt, Rh, Ir)[11] : Metalloporphyrins are prepared using an organic acid method as follows (Scheme 1-4 (a)). Metal-free porphyrin is refluxed with an excess metal chloride in an organic acid such as acetic acid at 100°C. The metalloporphyrin is obtained as a precipitate. The yield and purity of metalloporphyrin are very high.

(2)   Organic base method (Mg, Ca, Sr, Ba, Zn, Cd, Hg, Si, Ge, Sn, Pd, Ag, Au, Tl, As, Sb, Bi, Sc)[12, 13]: In this method, pyridine is usually used as the base. This method is widely used in the case of magnesium porphyrins. Metal-free porphyrin is refluxed with excess metal perchlorate in dry pyridine at 115 ~185°C for a few hours.

(3)   Acetylacetonate method (Mn, Fe, Co, Ni, Cu, Zn, Al, Sc, Ga, In, Cr, Mo, Ti, V, Zr, Hf, Eu, Pr, Yb, Y, Th)[14]: The reaction in this method is shown in Scheme 1-4 (b). The metal-free porphyrin is refluxed with metal acetylacetonate complex in organic solvent at 180~240°C for a few hours. Benzene, phenol and chlorobenzene are usually used as solvents in this method. Lanthanide porphyrins are widely synthesized using this method.

(4)   Metal carbonyl complex method (Cr, Mo, Mn, Tc, Re, Fe, Ru, Co, Rh, Ir, Ni, Os)[15,16] : The

reaction in this method is shown in Scheme 1-4 (c). In this reaction, metal is oxidized with the reduction of the pyrrol proton of porphyrin and then inserted into the porphyrin ring. Benzene and decalin are usually used as the solvents.

(5) Dimethylformamide method (Zn, Cu, Ni, Co, Fe, Mn, V, Hg, Cd, Pd, Sn, Mg, Ba, Ca, Pd, Ag, Rh, In, As, Sb, Tl, Bi, Cr)[17]: As many kinds of metal salts are dissolved in dimethylformamide (DMF) whose boiling point is rather high (153°C), the solvent is widely used in the synthesis of metalloporphyrins. Metalloporphyrins are synthesized by refluxing metal-free porphyrin with excess metal chloride at 153°C for a few hours.

(6) Benzonitrile method (Ta, Mo, W, Pd, Pt, Zr, In)[18]: As benzonitrile also can dissolve many kinds of metal salts with its high boiling point (191°C), the solvent is used in the synthesis of metalloporphyrins. Metalloporphyrins are synthesized by refluxing metal-free porphyrin with excess metal chloride at 191°C for a few hours.

The reaction process and completion are monitored using the UV-vis absorption spectrum of the Q-band of porphyrins. Purification of metalloporphyrins is performed by recrystallization or column chromatography. These metalloporphyrins are widely used as model compounds of artificial photosynthesis, photosensitizers, superconductors and photochromic materials.

1.1.6 Synthesis of chlorin and its derivatives [19]

Chlorin and its derivatives (called reduced porphyrin) have the same isoelectronic structure of the π-electron system of chlorophyll. *trans*-Octaethylchlorin is synthesized by the reduction of FeOEP as shown in Scheme 1-5. *trans*-Octaethylchlorin is purified using silica gel column

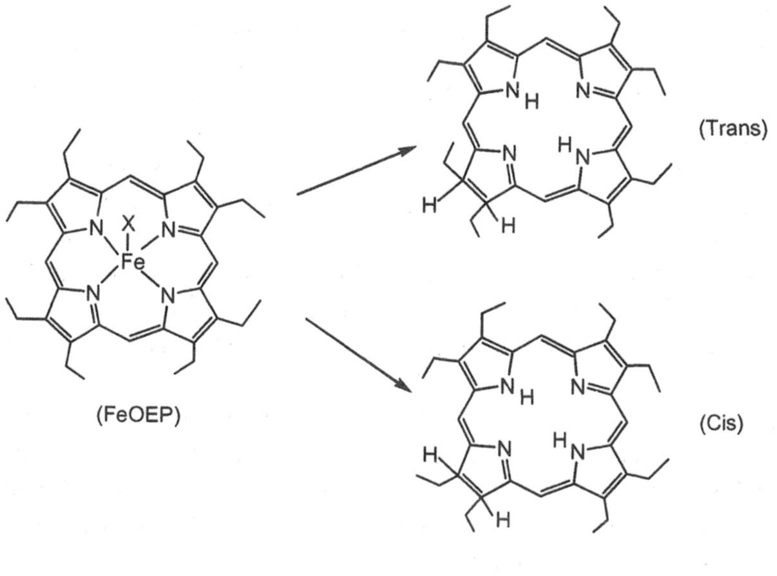

Scheme 1-5 Synthesis route of chlorin.

chromatography. *cis*-Octaethylchlorin is synthesized by the reduction of OEP in β-picoline and *p*-toluenesulfonylhydrazine under continuous nitrogen gas bubbling. As chlorin and its derivatives (metal complexes) are unstable under atmospheric conditions, these compounds are preserved under deoxygenated conditions using argon or inert gases.

1.2 General Methods for Preparation of Phthalocyanines

The general methods for the synthesis of phthalocyanines are described in this section. Since the early work by Linstead et al.,[20] many procedures for phthalocyanine preparation have been developed. These reactions are summarized in Scheme 1-6. Depending on the metal salt and the valence of the metal, a variety of phthalocyanines can be synthesized. The other methods are classified into four categories: (1) strong base method,[21] (2) lithium method, (3) 1,3-diiminoisoindoline method [22] and (4) electrolytic synthesis.[23-25]

(1) Strong base method: Phthalonitrile, 1,8-diazabicyclo[5,4,0]undensene (DBU) are refluxed with ethanol for 18-24 h.

(2) Lithium method: Lithium is dissolved in hot 1-pentanol then 1,4-dibutylthio-2,3-dicyanobenzene is added and the mixture is refluxed for 1 h. The precipitate is obtained by the addition of methanol with concentrated hydrochloric acid. The phthalocyanine is purified using silica gel column chromatography and recrystallized with chloroform-methanol.

(3) 1,3-Diiminoisoindoline method: 1,3-Diiminoisoindoline is refluxed with 1-diethylaminoethanol for 7 h. The precipitate is collected by suction filtration and washed with ethanol and acetone and dried under vacuum.

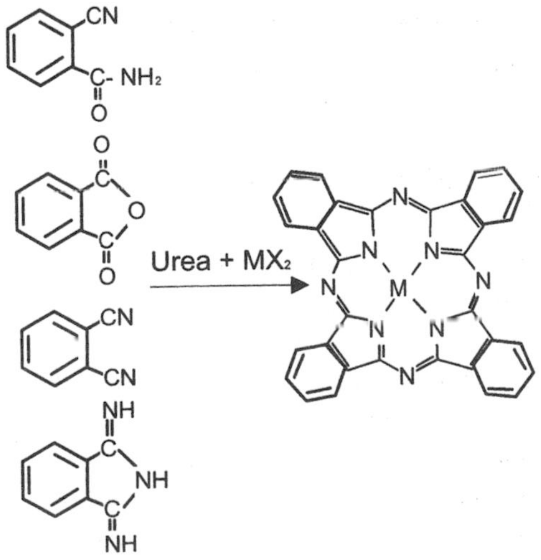

Scheme 1-6 Synthesis route of phthalocyanine.

(4) Electrolytic synthesis: Lithium chloride in anhydrous ethanol solution is added in the electrolytic cell and is then substituted by nitrogen gas. Phthalonitrile is added at the cathode side and electrolyzed at −1.6 V. After electrolysis, sulfonic acid is added to the reaction mixture to obtain the crude phthalocyanine as precipitate. The precipitate is washed with water and acetone then purified by Soxlet extraction with acetone.

1.2.1 Synthesis of water-soluble phthalocyanines

Tetrasulfophthalocyanine is synthesized by the following method.[26] Triammonium 4-sulfophthalate, dried urea, boric acid, ammonium molybdate and molybdane are mixed at 170-250°C for a few hours then extracted with water. After extraction, the precipitate is washed with ethanol and dissolved in ammonium carbonate saturated ammonia. Tetrasulfonatophthalocyanine ammonium salt is salted out. This salt is washed with ethanol then dissolved with hydrochloric acid and recrystallized with hydrochloric acid.

2,3,9,10,16,17,23,24-Octacarboxyphthalocyanine is synthesized by hydrolysis of octacyanophthalocyanine.[27] 1,2,4,5-Tetracyanobenzene is added to boiling 1-propanol then lithium in 1-propanol solution is added and refluxed for 20 min. After cooling, the precipitate is collected by suction filtration then washed with 1-propanol and ether to obtain octacyano-phthalocyanine. Octacyanophthalocyanine and potassium hydroxide are refluxed with triethylene glycol-water at 160°C for four days. After cooling, excess water is added to the reaction mixture and the brown filtrate is collected by suction filtration. When hydrochloric acid is added to the filtrate with stirring, the precipitate is obtained. The precipitate is washed with water (pH=5) then dried under vacuum to obtain octacarboxyphthalocyanine.

2,3,9,10,16,17,23,24-Octahydroxyphthalocyanine is synthesized by the cleavage of the ether bond of octamethoxyphthalocyanine.[28,29] Octamethoxyphthalocyanine is refluxed with pyridinium chloride for 30 min. After cooling, the precipitate is collected by suction filtration under nitrogen gas atmosphere then washed with water and acetone to obtain octahydroxy-phthalocyanine.

1.2.2 Synthesis of metallophthalocyanine

Metallophthalocyanines are also widely used as a model of metal enzymes and chlorophyll in the biochemical field and applied to functional optical devices. The standard methods of synthesis of metallophthalocyanines are described below.

(1) Alkaline and alkaline earth metal-phthalocyanines: Dilithium phthalocyanine is synthesized by the following method.[30] Lithium and phthalonitrile are refluxed with 1-pentanol for 30 min. After evaporation, the precipitate is dissolved in acetone and purified by Soxlet extraction with acetone. Dilithium phthalocyanine is obtained by recrystallization from hexane.

(2) Lanthanoids and actinoids-phthalocyanines: Mono- and bis-(phthalocyaninato) ruthethium (III) are synthesized by the following procedure.[31] Phthalonitrile, ruthethium acetate trihydrate

and DBU are refluxed with 1-hexanol for 5 h. After cooling, the precipitate is obtained by suction filtration then washed with acetic anhydride, cooled acetone and pentane and purified using silica gel column chromatography. Mono- and bis-(phthalocyaninato) ruthethium is eluted as the second and first fraction, respectively. Both compounds are obtained as precipitates by the addition of hexane.

(3) Superphthalocynanine (Fig. 1.1): Phthalonitrile and uranyl chloride anhydrate are refluxed with quinoline at 170 °C for 45 min under nitrogen atmosphere.[32] The precipitate is washed with methanol and acetone and purified by Soxlet extraction with anhydrate ethanol for 18 h. Superphthalocyanine is obtained by extraction from benzene solution.

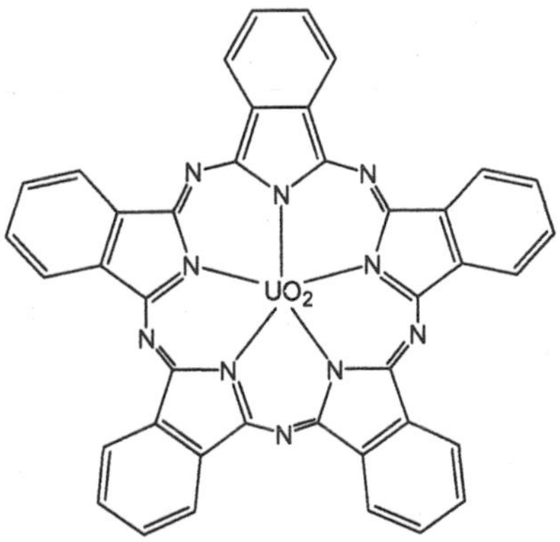

Fig. 1.1 Structure of superphthalocyanine.

(4) Titanium phthalocyanine: Oxotitanium phthalocyanine is synthesized as follows.[33] Phthalonitrile, titanium tetrabutoxide and urea are refluxed with 1-octanol at 150°C for 6 h then methanol is added and refluxed for 30 min. The precipitate is obtained by suction filtration then washed with toluene, methanol and water and dried at 100°C for 6 h to obtain oxotitanium phthalocyanine.

(5) Vanadium and ruthenium phthalocyanine: Vanadyl hexadecafluoro-phthalocyanine is prepared as follows.[34] Tetrafluorophthalonitrile and vanadium chloride are refluxed with 1-pentanol for 20 h and purified using silica gel column chromatography then dried under vacuum to obtain vanadyl hexadecafluoro-phthalocyanine.

Dibenzonitrile (phthalocyaninato) ruthenium (II) monohydrate is synthesized as follows.[35] Phthalonitrile, hydroquinone and ruthenium trichloride are refluxed with 1-pentanol for three days under ammonia gas atmosphere. After cooling, the precipitate is washed with methanol and dichloromethane to obtain bis-ammonia (phthalocyaninato) ruthenium (II). Bis-ammonia (phthalocyaninato) ruthenium (II) is refluxed with benzonitrile for one day under nitrogen atmosphere. After suction filtration, dibenzonitrile (phthalocyaninato) ruthenium (II) monohydrate is obtained by the addition of excess methanol to the filtrate.

(6) Subphthalocyanine (Fig. 1.2)[36] : Phthalonitrile is refluxed with 1-chloronaphthalene then boron trichloride is added and the mixture is refluxed for 10 min. After cooling, subphthalocyanine is obtained as precipitate.

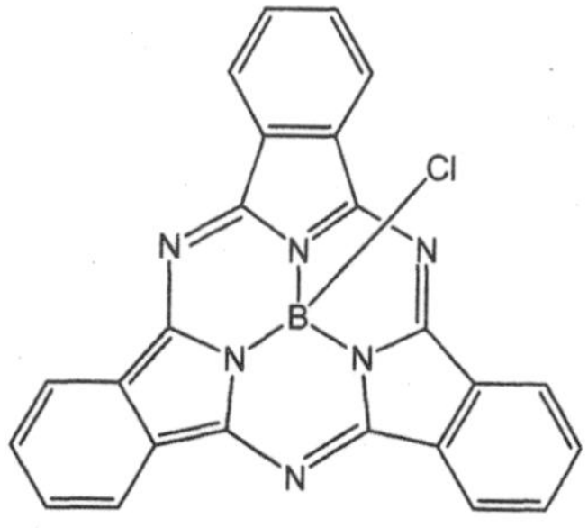

Fig. 1.2 Structure of subphthalocyanine.

(7) Aluminum phthalocyanine[37] : Tetra-*tert*-butylphthalocyaninatoaluminum hydroxide is obtained by the following method. Tetra-*tert*-butylphthalocyanine and aluminum chloride are refluxed with quinoline at 150°C for 4 h. After cooling, excess chloroform is added to the reaction mixture and the filtrate is collected by suction filtration. After evaporation, the precipitate is obtained by the addition of hydrochloric acid and methanol. Tetra-*tert*-butylphthalocyaninato-aluminum hydroxide is purified using silica gel column chromatography.

(8) Indium phthalocyanine: Bis(phthalocyaninato) indium is synthesized by the following method.[38] Indium-magnesium alloy and phthalonitrile are mixed and heated in a deoxygenated sample at 210°C for 24 h. Bis(phthalocyaninato) indium is obtained as blue-purple crystals.

(9) Antimonate phthalocyanine: Dichloro(phthalocyaninato) antimonate is prepared as follows.[39] Antimony trichloride and phthalonitrile is heated at 150°C for 10 h under argon atmosphere. The crude sample is dissolved with dichloromethane then extracted repeatedly. Dichloro(phthalo-cyaninato) antimonate is obtained as dark-blue crystals.

1.3   New Porphyrins and Phthalocyanines

1.3.1 Viologen-linked porphyrins

Many types of viologen-linked porphyrins have been synthesized. In viologen-linked porphyrins, photosensitizer (porphyrin) and electron carrier (viologen) are conjugated and they serve as photosensitizer and electron carrier in the same molecule. The syntheses of typical viologen-

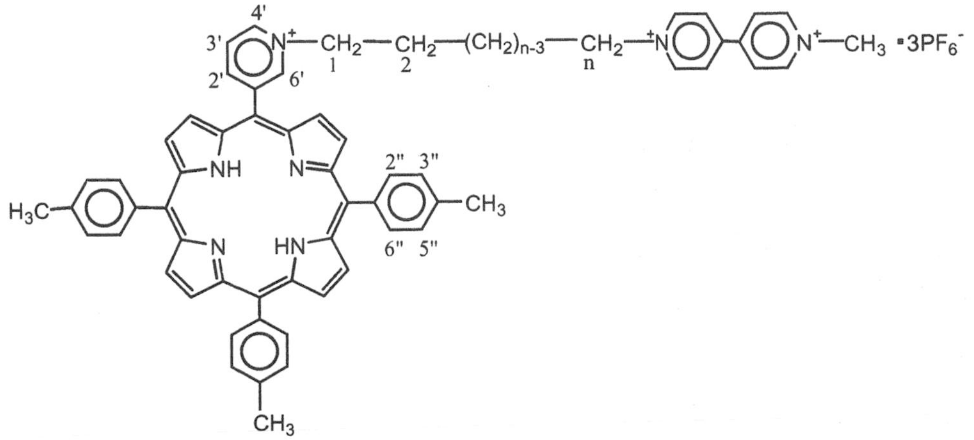

Fig. 1.3   Structure of viologen-linked porphyrin (*m*-PC$_n$V). (Reproduced with permission by C. Franco, C. McLendon, *Inorg. Chem.*, **23** (1984) 2370 , ACS)

linked porphyrins (Fig. 1.3) are described.[40] The synthesis route is shown in Scheme 1-7. The starting material, 5-(4-pyridyl)-10,15,20-tritolylphenylporphyrin (PyTP), is synthesized by the following method.[41] Pyridine-4-aldehyde and *p*-tolylbenzaldehyde are added to boiling propionic acid, followed by pyrrol, and the mixture is refluxed at 165°C for 1 h. Metallic purple precipitate is collected by suction filtration. The crude product obtained is purified by silica gel column chromatography. The desired product, PyTP, is eluted as a second fraction. PyTP is quaternized with an excess amount of α,ω-dibromoalkane (BrC$_n$Br) (n=3-6) in toluene at 110°C for 48 h to obtain 5-(4-bromoalkyl- pyridinium)-10,15,20-tritolyl-phenylporphyrin (TPPC$_n$Br). TPPC$_n$Br and an excess amount of 1-methyl-4, 4′-bipyridinium are refluxed in dimethylformamide (DMF) for 96 h to obtain viologen-linked porphyrin (PC$_n$V). The desired product is obtained by gel column chromatography. As an example, 500 MHz ^{1}H-NMR spectrum of PC$_4$V in acetonitrile-d_3 is shown in Fig. 1.4. The absorption spectrum of PC$_4$V is shown in Fig. 1.5 as an example. All the absorption spectra of PC$_n$V are similar to that of viologen-free porphyrin, indicating the absence of any ground state electronic interaction between the porphyrin ring and the bonded viologen.

Bis-viologen-linked porphyrin (Fig. 1.6) is prepared as shown in Scheme 1-8.[42] PyTP and an excess amount of 1-bromobutyl-1′-(n-butyl)-4,4′-bipyridinium are refluxed in toluene and methanol solution for 96 h to obtain bis-viologen-linked porphyrin (PC$_4$V$_A$C$_4$V$_B$). The desired product is obtained by gel column chromatography. 200 MHz ^{1}H-NMR spectrum of PC$_4$V$_A$C$_4$V$_B$ in DMSO-d_6 is shown in Fig. 1.7. The absorption spectrum of PC$_4$V$_A$C$_4$V$_B$ is shown in Fig. 1.8. The absorption spectrum of PC$_4$V$_A$C$_4$V$_B$ is similar to that of viologen-free porphyrin, indicating the absence of any ground state electronic interaction between the porphyrin ring and the bonded viologens.

Water-soluble viologen-linked porphyrins and bis-viologen-linked porphyrin have been

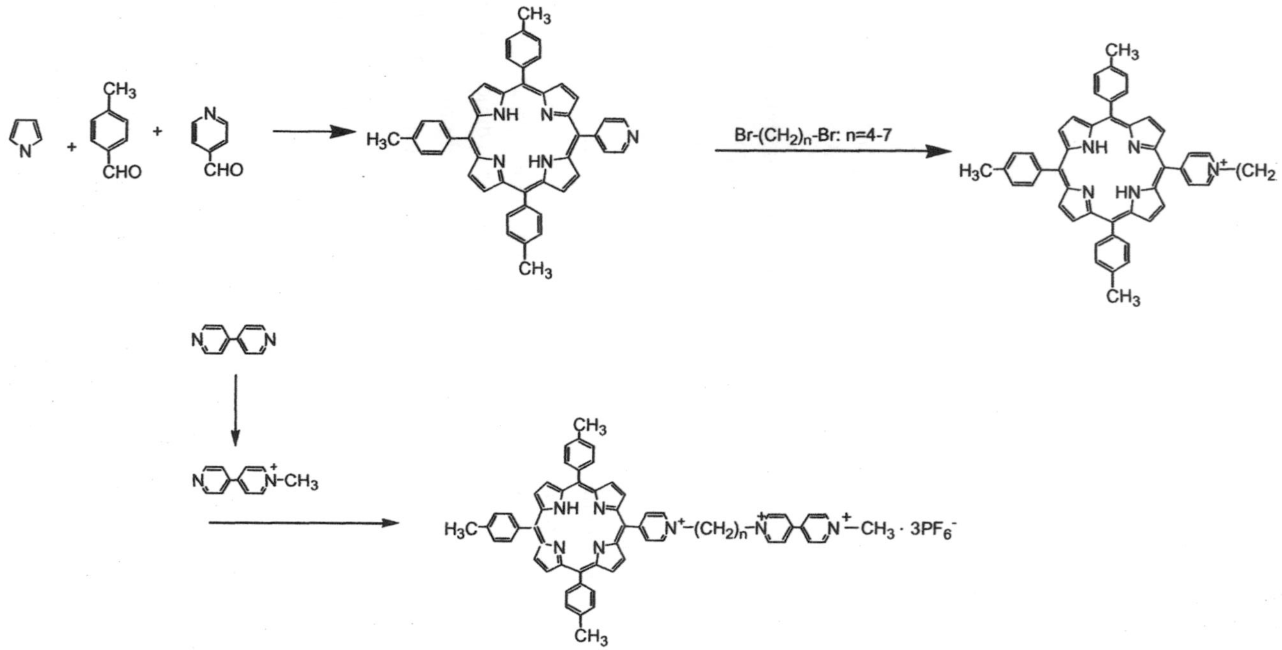

14

Scheme 1-7 Synthesis route of viologen-linked porphyrin.

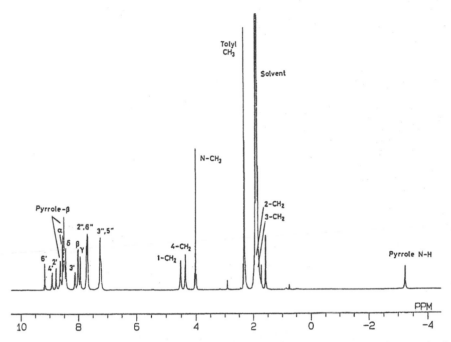

Fig. 1.4   500 MHz ^{1}H NMR spectrum of m-PC$_4$V in MeCN-d_3. The signal assignment is shown in the spectrum. (Reproduced with permission by Y. Yamamoto et al., *Bull. Chem. Soc. Jpn.*, **62** (1989) 2152)

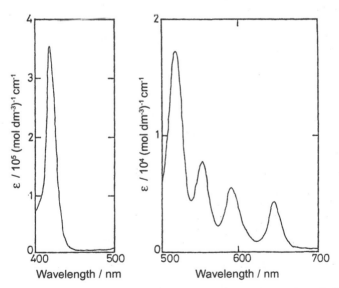

Fig. 1.5   Absorption spectrum of m-PC$_4$V. (Reproduced with permission from *Bull. Chem. Soc. Jpn.*, **62** (1989) 2152)

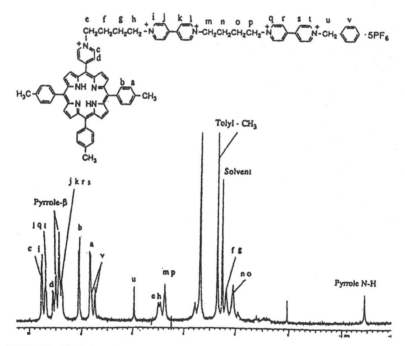

Fig. 1.6   Structure of bis-viologen-linked porphyrin (PC$_4$V$_A$C$_4$V$_B$). (Reproduced with permission from *J. Photochem. Photobiol. A: Chem.*, **77** (1994) 29 , Elsevier Science)

Fig. 1.7   200 MHz ^{1}H NMR spectrum of PC$_4$V$_A$C$_4$V$_B$ in DMSO-d_6. The signal assignment is shown in the spectrum. (Reproduced with permission from *J. Photochem. Photobiol. A: Chem.*, **77** (1994) 29 , Elsevier Science)

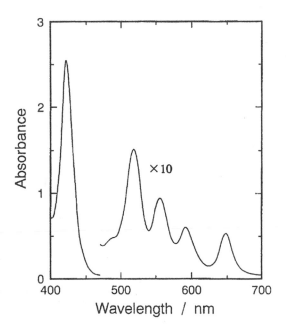

Fig. 1.8   Absorption spectrum of $PC_4V_AC_4V_B$. (Reproduced with permission from *J. Photochem. Photobiol. A: Chem.*, **77** (1994) 29, Elsevier Science)

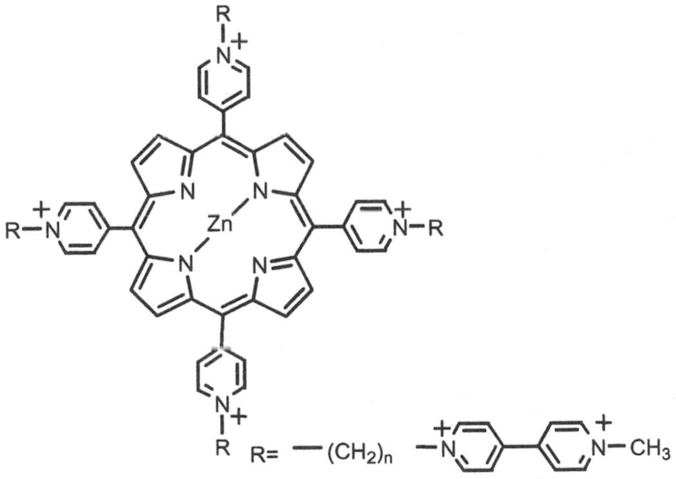

Fig. 1.9   Structure of water-soluble viologen-linked zinc cationic porphyrin ($ZnP(C_nV)_4$). (Reproduced with permission from *J. Photochem. Photobiol. A: Chem.*, **77** (1994) 29, Elsevier Science)

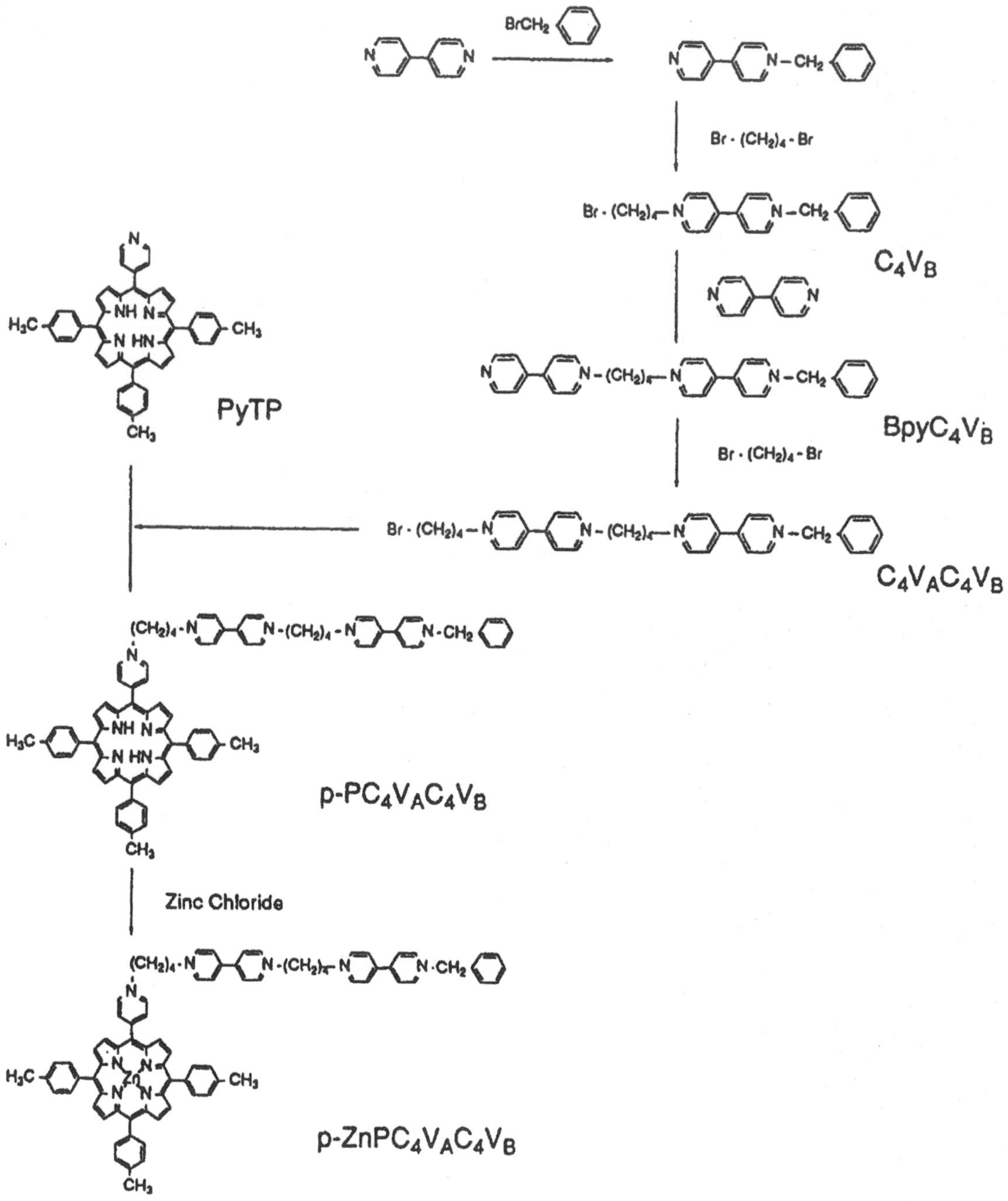

Scheme 1-8   Synthesis route of bis-viologen-linked porphyrin. (Reproduced with permission from *J. Photochem. Photobiol. A: Chem.*, **77** (1994) 29, Elsevier Science)

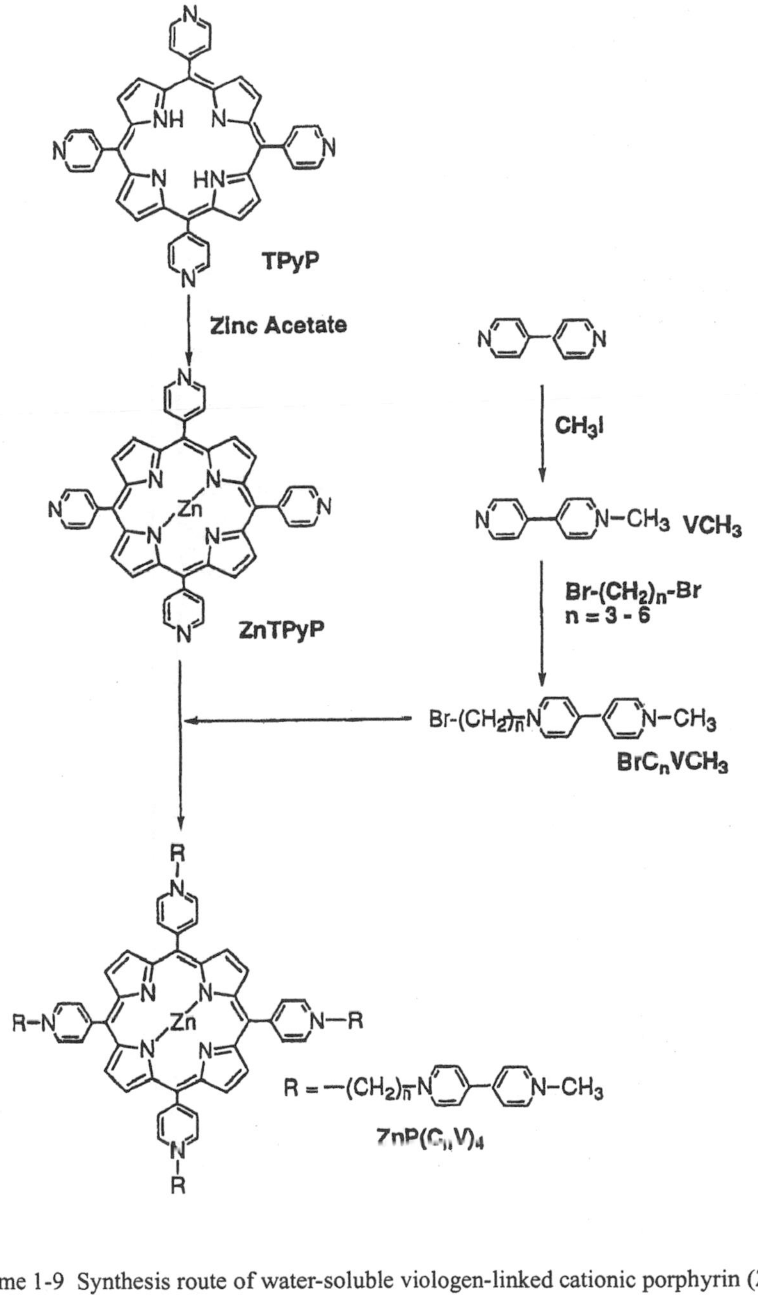

Scheme 1-9 Synthesis route of water-soluble viologen-linked cationic porphyrin $(ZnP(C_nV)_4)$.

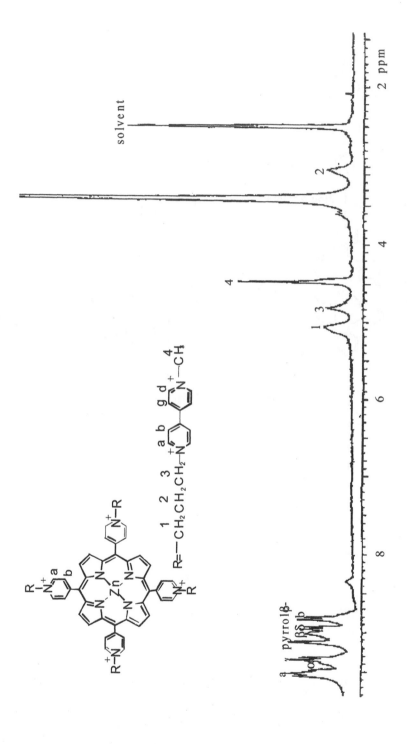

Fig. 1.10 200 MHz ^{1}H NMR spectrum of ZnP(C$_n$V)$_4$ in DMSO-d_6. The signal assignment is shown in the spectrum. (Reproduced with permission from *J. Photochem. Photobiol. A: Chem.*, **98** (1996) 59, Elsevier Science)

synthesized. Water-soluble viologen-linked cationic porphyrin (ZnP(C$_n$V)$_4$, Fig. 1.9) is synthesized as the following Scheme 1-9.[43] Zinc tetrakis (4-pyridyl) porphyrin (ZnTPyP) and an excess amount of 1-bromoalkyl-1'-methyl-4,4'-bipyridinium are refluxed in DMF for 96 h to obtain ZnP(C$_n$V)$_4$. The desired product is obtained by gel column chromatography. As an example, 200 MHz ^{1}H-NMR spectrum of ZnP(C$_3$V)$_4$ in DMSO-d_3 is shown in Fig. 1.10. The absorption spectrum of ZnP(C$_3$V)$_4$ is shown in Fig. 1.11 as an example.

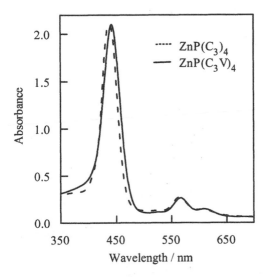

Fig. 1.11 Absorption spectra of ZnP(C$_3$V)$_4$ and ZnP(C$_3$)$_4$. (Reproduced with permission from *J. Photochem. Photobiol. A: Chem.*, **98** (1996) 59, Elsevier Science)

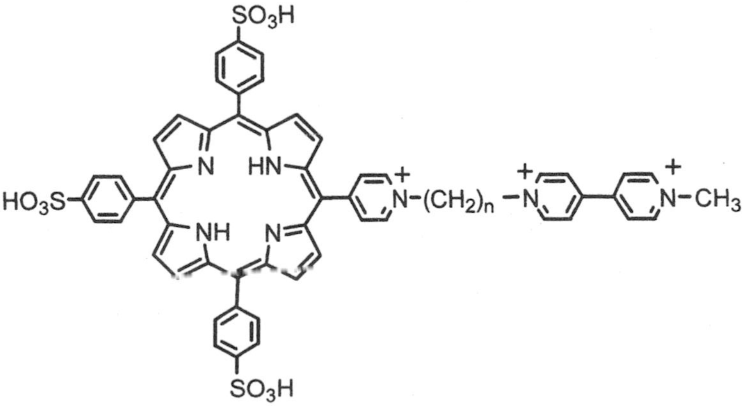

Fig. 1.12 Structure of water-soluble viologen-linked anionic porphyrin (TPPSC$_n$V).

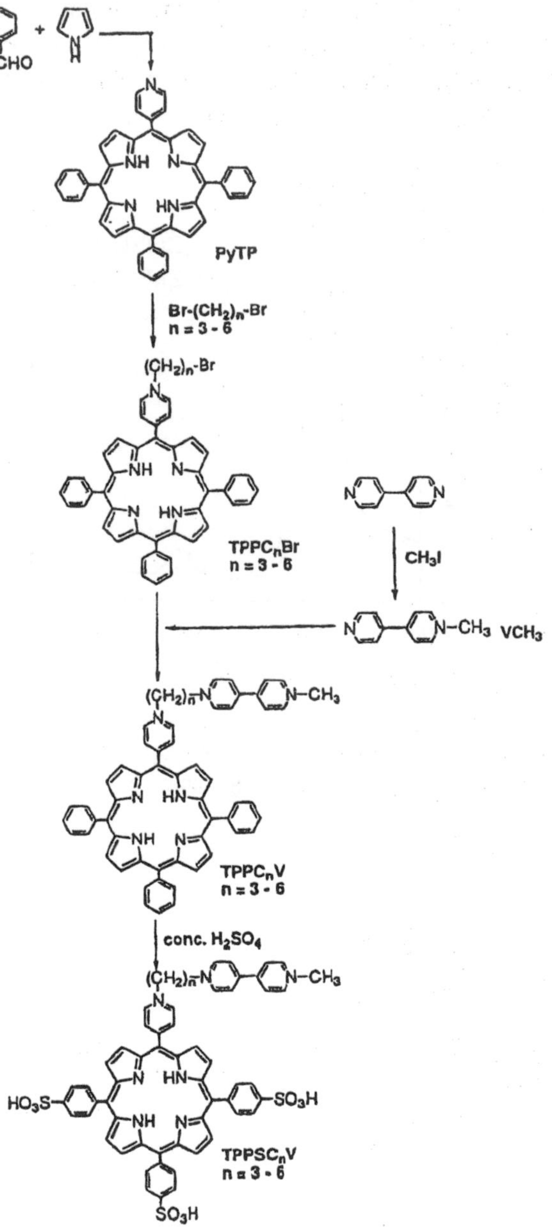

Scheme 1-10 Synthesis route of water-soluble viologen linked anionic porphyrin (TPPSC$_n$V).

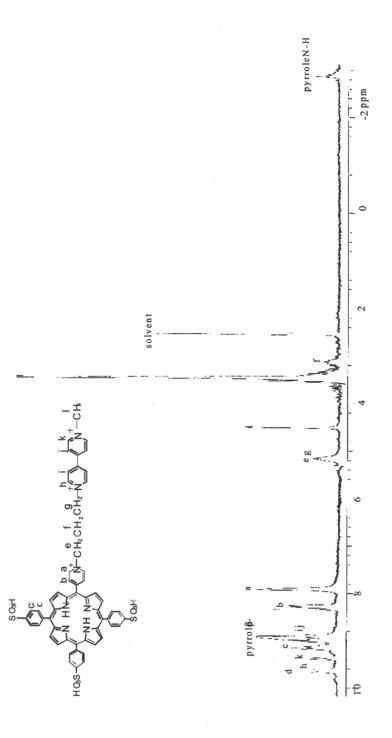

Fig. 1.13 200 MHz ¹H NMR spectrum of TPPSC$_n$V in DMSO-d_6. The signal assignment is shown in the spectrum.(Reproduced with permission from *J. Mol. Cctal., A: Chem.*, **126** (1997)13, Elsevier Science)

The synthesis route of water soluble viologen-linked anionic porphyrin (TPPSC$_n$V, Fig. 1.12) is shown in Scheme 1-10. TPPSC$_n$V is synthesized by sulfonation of viologen-linked triphenylporphyrin (TPPC$_n$V).[44] The sulfonation is performed by the following procedure. TPPC$_n$V is dissolved in 20 ml of conc. H$_2$SO$_4$ and is heated to reflux for 4 h and the reaction mixture is stirred at room temperature for 48 h. The reaction mixture is diluted with a double volume of water and excess acetone is added to the reaction mixture and green precipitate is collected by suction filtration. As an example, 200 MHz ^{1}H-NMR spectrum of TPPSC$_3$V in DMSO-d_3 is shown in Fig. 1.13. The absorption spectrum of TPPSC$_3$V is shown in Fig. 1.14 as an example. All the absorption spectra of TPPSC$_n$V are similar to that of viologen-free porphyrin, indicating the absence of any ground state electronic interaction between the porphyrin ring and the bonded viologen.

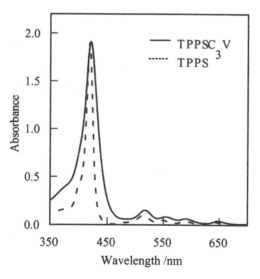

Fig. 1.14 Absorption spectra of TPPSC$_3$V and TPPS. (Reproduced with permission from *J. Mol. Catal., A: Chem.,* **126** (1997)13 , Elsevier Science)

1.3.2 Quinone-linked porphyrins

Quinone-linked porphyrins are synthesized as model compounds of the photosynthesis reaction center and their photoexcited intramolecular electron transfer processes are reported. The syntheses of typical quinone-linked porphyrins (Fig. 1.15) are described.[45] The synthesis route is shown in Scheme 1-11. The Wittig reaction of **3** with triphenylphosphoran **4**, prepared from the corresponding phosphonium salt and butyl lithium in THF, is carried out at room temperature to give **5**. Hydrogenation of **5** fails due to steric hindrance around the double bonds. Compound **6** obtained by treatment with tribromo borane in dichloromethane at −20°C is smoothly

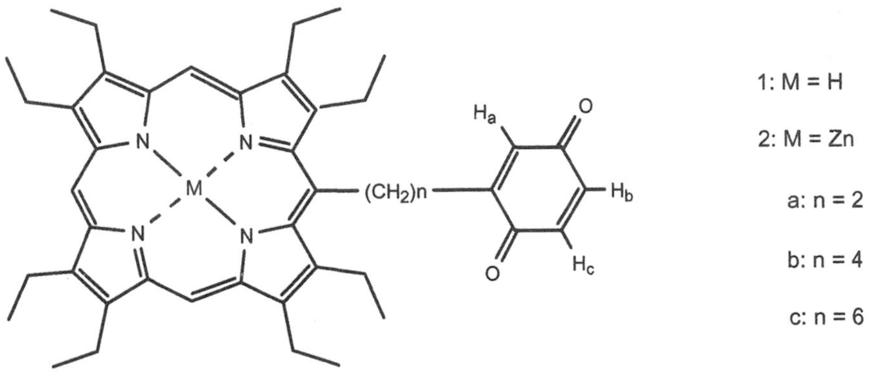

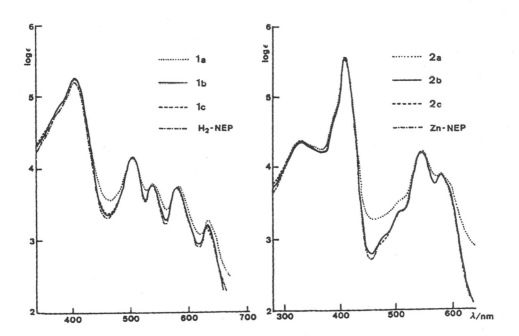

Fig. 1.15 Structure of quinone-linked porphyrins. (Reproduced with permission by S. Nishitani et al., *Tetrahedron Lett.,* **22** (1981) 2099 , Elsevier Science)

Fig. 1.16 Absorption spectra of metal-free complexes of Fig.1.15 in THF. (Reproduced with permission by S. Nishitani et al., *Tetrahedron Lett.,* **22** (1981) 2099 , Elsevier Science)

Fig. 1.17 Absorption spectra of zinc complexes of Fig.1.15 in THF. (Reproduced with permission by S. Nishitani et al., *Tetrahedron Lett.,* **22** (1981) 2099 , Elsevier Science)

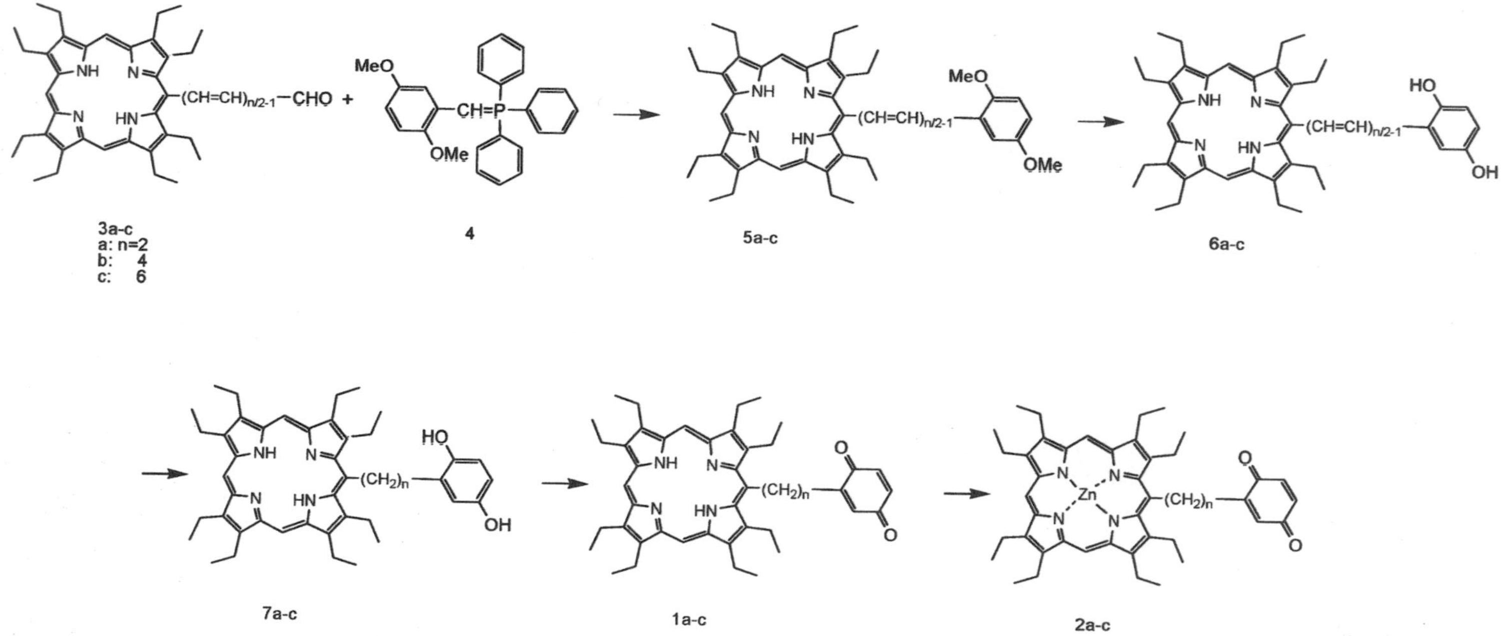

Scheme 1-11 Synthesis route of quinone-linked porphyrin.

hydrogenated with 10% Pd-C in benzene to afford **7**. Oxidation of **7** with DDQ in benzene gives the desired compound **1**. Treatment of **1** with zinc acetate in chloroform under reflux for 10 min yields zinc complex 2. The absorption spectra of **1a-c** and **2a-c** are shown together with that of nonaethylporphyrin (NEP) as the reference compound in Figs. 1.16 and 1.17. All the spectra are similar to that of NEP or ZnNEP, indicating no appreciable interaction between the porphyrin and quinone in the ground state. Relative fluorescence intensities of **1**, **2** and NEP in the solvents with different polarity are summarized in Table 1.1. The most characteristic feature is that the fluorescence intensities of all the quinone-linked porphyrins decrease as much as those of metal-free NEP and ZnNEP. To estimate the efficiency of such intramolecular quenching, relative quenching efficiencies are calculated by comparing the quenching constant ($k_f \tau_p$), evaluated by the Stern-Volmer plot for the fluorescence intensities of metal-free NEP or ZnNEP and toluquinone mixtures. The electron transfer occurs very efficiently in the intramolecular systems where porphyrin and quinone are closely situated and the coordinated metal as well as the shortening of the polymethylene chain make the quenching efficiency larger.

One of the model compounds for multistep electron transfer using quinone-linked porphyrins is shown in Fig. 1.18 (P4Q4Q′), where two types of quinone rings are connected to etioporphyrin so as to produce a redox potential gradient in the molecule.[46] The synthesis route is shown in Scheme 1-12. By the Wittig reaction of 2,5-dimethoxycinnamaldehyde and

Table 1.1   Relative fluorescence intensities of **1, 2**, and M-NEP

	CH*	THF	PrCN		CH*	THF	PrCN
H₂-NEP	385	397	336	Zn-NEP	541	487	351
1a	171	2	1	**2a**	12	3	2
1b	21	5	4	**2b**	9	2	1
1c	46	30	24	**2c**	23	11	8

* cyclohexane

(Reproduced with permission by S. Nishitani et al., *J. Am. Chem. Soc.*, **105** (1983) 7771, ACS)

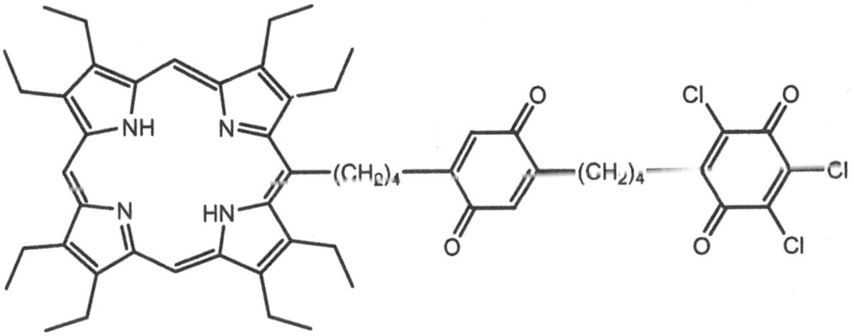

Fig. 1.18 Structure of quinone-linked porphyrin (P4Q4Q′). (Reproduced with permission by S. Nishitani et al., *J. Am. Chem. Soc.*, **105** (1983) 7771, ACS)

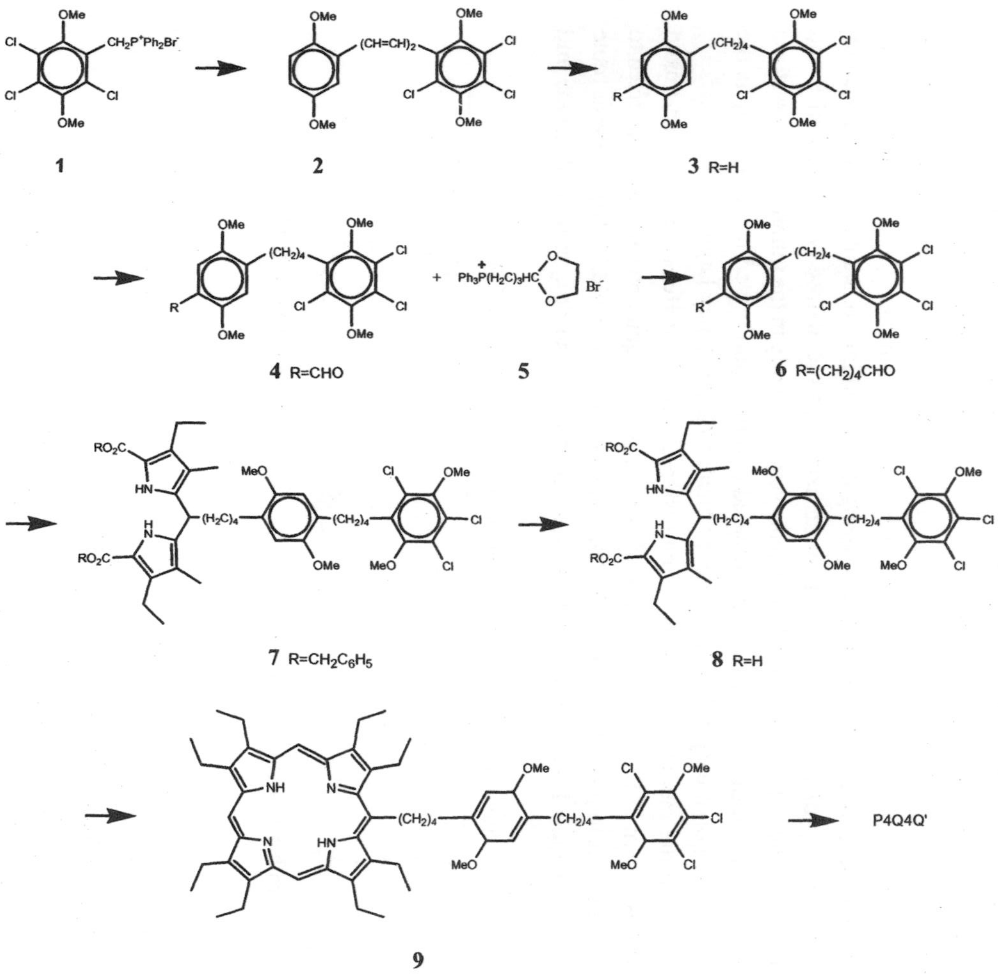

28

Scheme 1-12 Synthesis route of quinone-linked porphyrin (P4Q4Q').

phosphonium bromide **1**, prepared via two steps from 3,4,6-trichloro-2,5-dimethoxytoluene, butadiene **2** is given quantitatively, which is converted to **3** by hydrogenation over Pd-C. By the Rieche reaction of **3** is carried out with dichloromethyl methyl ether and titanium tetrachloride in dichloromethane at room temperature aldehyde **4** is given in 74% yield. Condensation of **4** with phosphonium salt **5** followed by hydrogenation over Pd-C and by deprotection with acid gives **6** in 71% yield. Aldehyde **6** is condensed with benzyl 3-ethyl-4-methylpyrrole-2-carboxylate in the presence of p-TsOH to give **7** in 85% yield. Catalytic hydrogenation of **7** over 10% Pd-C gives quantitatively **8**, which is coupled with 4,4′-diethyl-5,5′-diformyl-3,3′-dimethyl-2,2′-methylenebis(pyrrole) in a manner similar to that reported by Collman et al.[47] to give **9** in 24% yield. Demethylation of **9** with 30-fold of tribromoborate at –20°C for 2 h and then at room temperature for 5 h followed by oxidation with DDQ gives the desired compound P4Q4Q′ in 24% yield. The absorption spectrum of P4Q4Q′ in the region of 350 – 650 nm is very similar to that of 5-ethyl-etioporphyrin (EEP), indicating no significant electronic interaction among the porphyrin and two different quinones in the ground state. Moreover, the redox potentials (in acetonitrile vs. SCE) of the two quinones incorporated in P4Q4Q′(–0.75, –0.22 V) are almost the same as those of the reference quinones (2,5-dimethyl-p-benzoquinone, –0.73 V; trichlorotoluquinone, –0.24 V). This fact also suggests negligible interaction among the three chromophores and the existence of the gradient redox potential in the molecule.

1.3.3 Pyromellitimide-linked porphyrins

Pyromellitimide derivatives have been studied as effective electron carriers. As the reduced pyromellitimide has the absorption band in the visible region, it is easy to monitor the reaction rate by the absorption of the reduced pyromellitimide.

The syntheses of typical pyromellitimide-quinone-linked porphyrins (Fig. 1.19) are described.[48] The synthesis route is shown in Scheme 1-13. Compound **2a** is synthesized as follows. The aromatic aldehyde **1a**, bis-(3-hexyl-4-methyl-2-pyrrolyl)methane, and p-tolualdehyde are dissolved in chloroform-benzene solution under nitrogen. To this solution is added trichloroacetic acid in acetonitrile, and the resulting mixture is stirred for 16 h at room temperature in the dark. p-Chloranil is then added, and the resulting mixture is stirred again for 3 h. The solvent is evaporated then roughly purified by alumina column with dichloromethane as an elutant. After metalation with zinc acetate, zinc porphyrin is separated by flash column

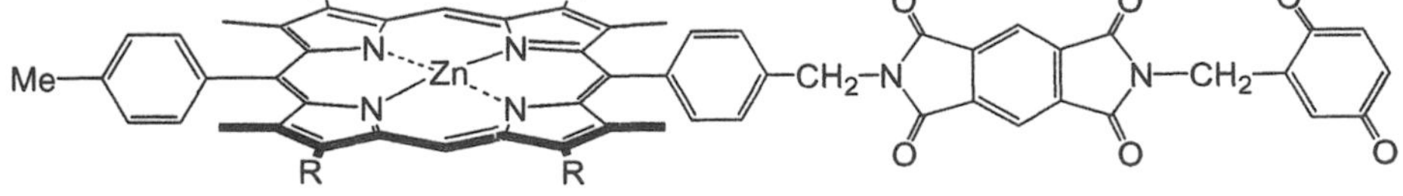

Fig. 1.19 Structure of Pyromellitimide-quinone-linked porphyrin. (Reproduced with permission by A. Takahashi et al., *J. Am. Chem. Soc.,* **115** (1993) 12137, ACS)

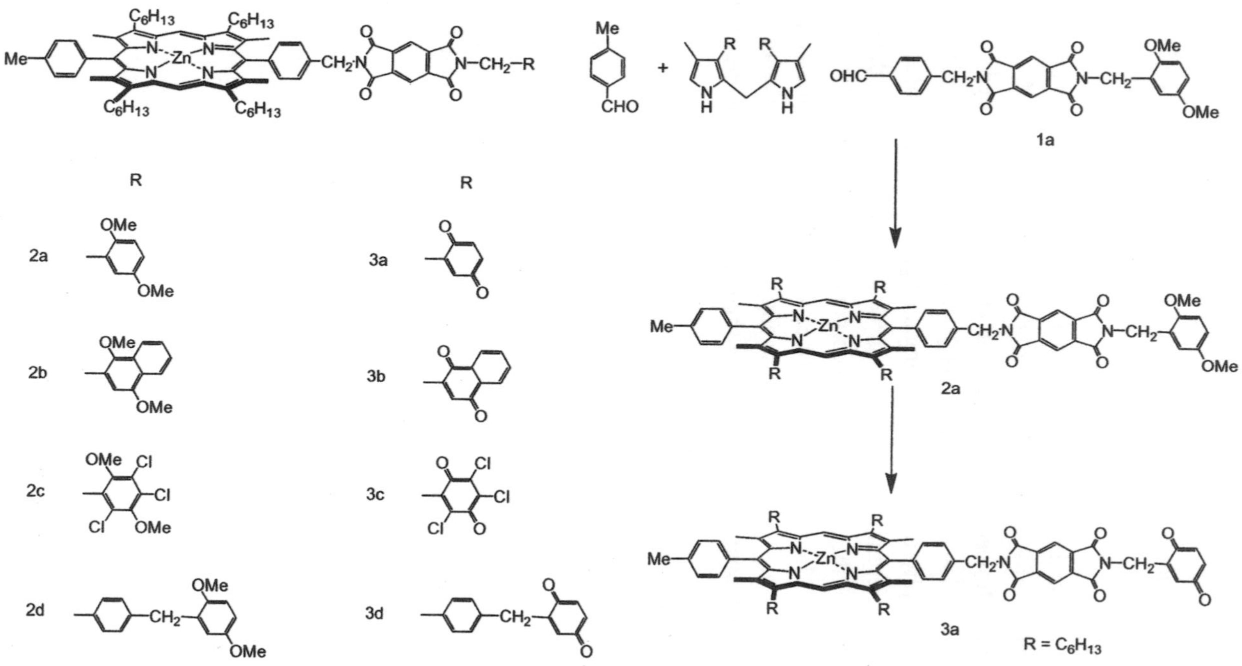

Scheme 1-13 Synthesis route of pyromellitimide-quinone-linked porphyrin. (Reproduced with permission by A. Takahashi et al., *J. Am. Chem. Soc.*, **115** (1993) 12137, ACS)

chromatography (silica gel, dichloromethane). Recrystallization from dichloromethane-*n*-hexane gives **2a** in 18% yield. Compounds **2b-2d** are prepared in a similar manner. Compound **3a** is synthesized as follows. Compound **2a** is dissolved in dry dichloromethane, and the solution is cooled to −78°C under nitrogen. To this solution is added tribromoborate in dichloromethane slowly in the dark. The reaction mixture is stirred for 1 h at the same temperature and is later allowed to warm to room temperature. The reaction mixture is then cooled to 0°C, and water is added carefully to quench the excess tribromoborate. The organic layer is separated, and the aqueous layer is further extracted with dichloromethane. The organic layers are combined, washed with water and aqueous sodium bicarbonate, dried over anhydrous sodium sulfate, and taken to dryness on a rotary evaporator. The hydroquinone-linked porphyrin product is quickly purified by flash column chromatography (silica gel, dichloromethane), dissolved in dichloromethane and oxidized with PbO_2 under nitrogen. The reaction mixture is stirred for 2 h in the dark at room temperature. PbO_2 is filtered off, and the filtrate is treated with zinc acetate. Compound 3a is purified by flash column chromatography (silica gel, dichloromethane). Recrystallization from dichloromethane-*n*-hexane gives **3a** in 65% yield. Compounds **3b-3d** are prepared in a similar manner.

1.3.4 Carotene-linked porphyrins

One respect in which the carotene-linked porphyrins differ from natural reaction centers is that the photoinduced electron transfer step involves porphyrin-quinone transfer, whereas in reaction centers, the initial transfer is to a pheophytin (a chlorophyll lacking a central metal atom). Functional acceptor-linked porphyrins featuring interporphyrin photoinitiated electron transfer can be prepared if the redox potentials of the porphyrin moieties are adjusted so that there is sufficient driving force. Here carotene-linked porphyrin with a metal-free porphyrin with three pentafluorophenyl groups is introduced. The syntheses of typical carotene-linked porphyrins (Fig. 1.20) are described.[49] The synthesis route is shown in Scheme 1-14. 5-(4-Carbomethoxyphenyl)-10,15,20-tris-(pentafluorophenyl)porphyrin is prepared via a modification of the porphyrin synthesis developed by Lindsey.[50] Pentafluoroaldehyde, pyrrole and methyl 4-formylbenzoate are dissolved in trifluoroacetic acid and chloroform then refluxed for 4.5 h. After cooling, 2,3-dichloro-5,6-dicyano-1,4-benzoquinone is added to the mixture and stirred overnight. The reaction mixture is neutralized with sodium bicarbonate and extracted with four 100-ml portions of dichloromethane. The organic layer is separated, and the solvent is removed *in vacuo*. After the solution is filtered through a short silica gel column to remove tars, the porphyrins are separated by chromatography on silica gel with toluene elution. Recrystallization from dichloromethane and methanol gives pure 5-(4-carbomethoxyphenyl)-10,15,20-tris(pentafluorophenyl)porphyrin in 1.49% yield. Carotene-linked diporphyrin (1) is synthesized as follows. 5-(4-Carbomethoxyphenyl)-10,15,20-tris(pentafluorophenyl)porphyrin is hydrolyzed to its acid by refluxing in HCl and trifluoroacetic acid for 4 days, neutralizing to approximately pH 7 with aqueous NaOH, and extracting with chloroform. The chloroform is removed by

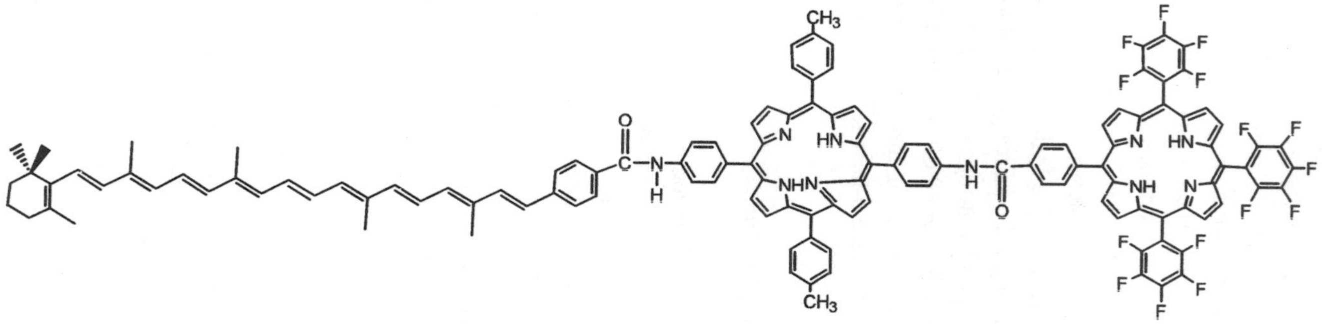

Fig. 1.20 Structure of carotene-linked porphyrin. (Reproduced with permission by D. Gust et al., *J. Am. Chem. Soc.*, **113** (1991) 3638, ACS)

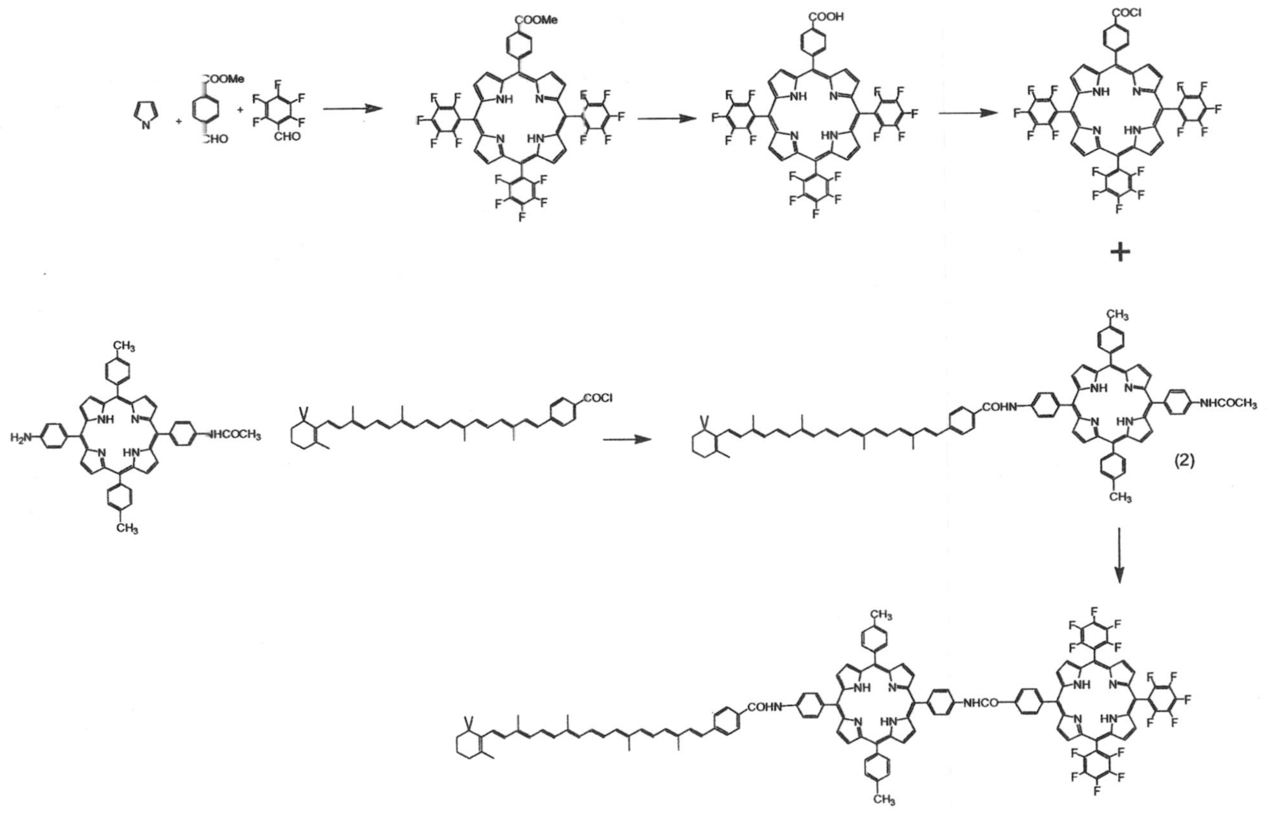

Scheme 1-14 Synthesis route of carotene-linked porphyrin. (Reproduced with permission by D. Gust et al., *J. Am. Chem. Soc.*, **113** (1991) 3638, ACS)

distillation at reduced pressure to yield the pure acid. 5-(4-Carboxyphenyl)-10,15,20-tris(pentafluorophenyl)porphyrin is obtained in 96% yield. 5-(4-Carboxyphenyl)-10,15,20-tris(pentafluorophenyl)porphyrin and toluene, pyridine and thionyl chloride are stirred for 20 min before evaporating the solvent at reduced pressure. Toluene is added to the residue, and the solvent is again evaporated under reduced pressure to remove all traces of thionyl chloride. The resulting acid chloride 5-(4-chlorocarboxyphenyl)-10,15,20-tris(pentafluorophenyl)porphyrin is dissolved in toluene, and the solution is added to a solution of (2) synthesized by coupling of 5-(4-aminophenyl)-15- (4-acetoamidephenyl)-10.20-bis(4-methylphenyl)porphyrin and carotene acid chloride, dissolved in a mixture of toluene and pyridine. The reaction mixture is stirred for 13 h under nitrogen before evaporation of solvent at reduced pressure. The resulting residue is dissolved in a mixture of water and dichloromethane and extracted with dichloromethane. The organic layer is evaporated to dryness and dissolved in toluene, and the toluene is distilled to remove any residual water and pyridine. The residue is chromatographed on silica gel (2% acetone in dichloromethane) to yield 62% pure carotene-linked diporphyrin.

1.3.5 Fullerene-linked porphyrins

Fullerene has unique electrochemical and photophysical properties both as an electron acceptor and a photosensitizer. Fullerene has moderate electron-accepting abilities similar to those of benzoquinones and naphthoquinones and reversibly accepts up to six electrons, and the energy levels of the first excited singlet state and triplet state are comparable to those of porphyrins, and are much lower than those of acceptors such as quinones and pyromellitic diimides.[51-57] The syntheses of typical fullerene-linked porphyrins (Fig. 1.21) are described.[58-60] The synthesis route is shown in Scheme 1-15. Dibromide **2** is obtained by coupling reaction of the corresponding aminoporphyrin with 3,4-bis(bromomethyl)benzoic acid in the presence of 2-chloro-4,6-dimethoxy-1,3,5-triazine and N-methylmorpholine in THF in 24% yield. The Diels-Alder reaction of **2** and C_{60} using potassium iodide and 18-crown-6 in toluene give **1a** in 50% yield. Reference

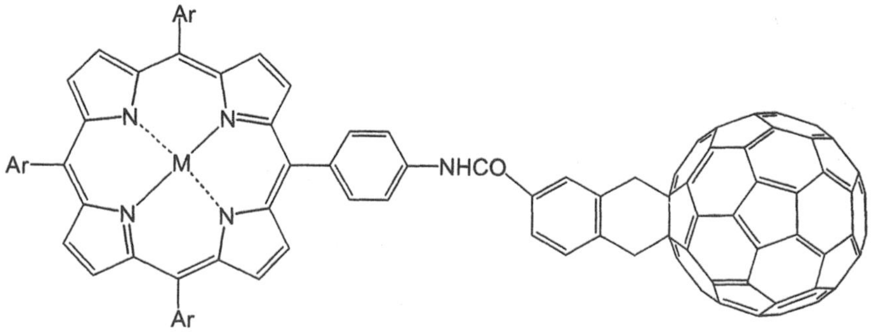

Fig. 1.21 Structure of fullerene-linked porphyrin. (Reproduced with permission by H. Imahori et al., *Chem. Lett.,*(**1995**) 265)

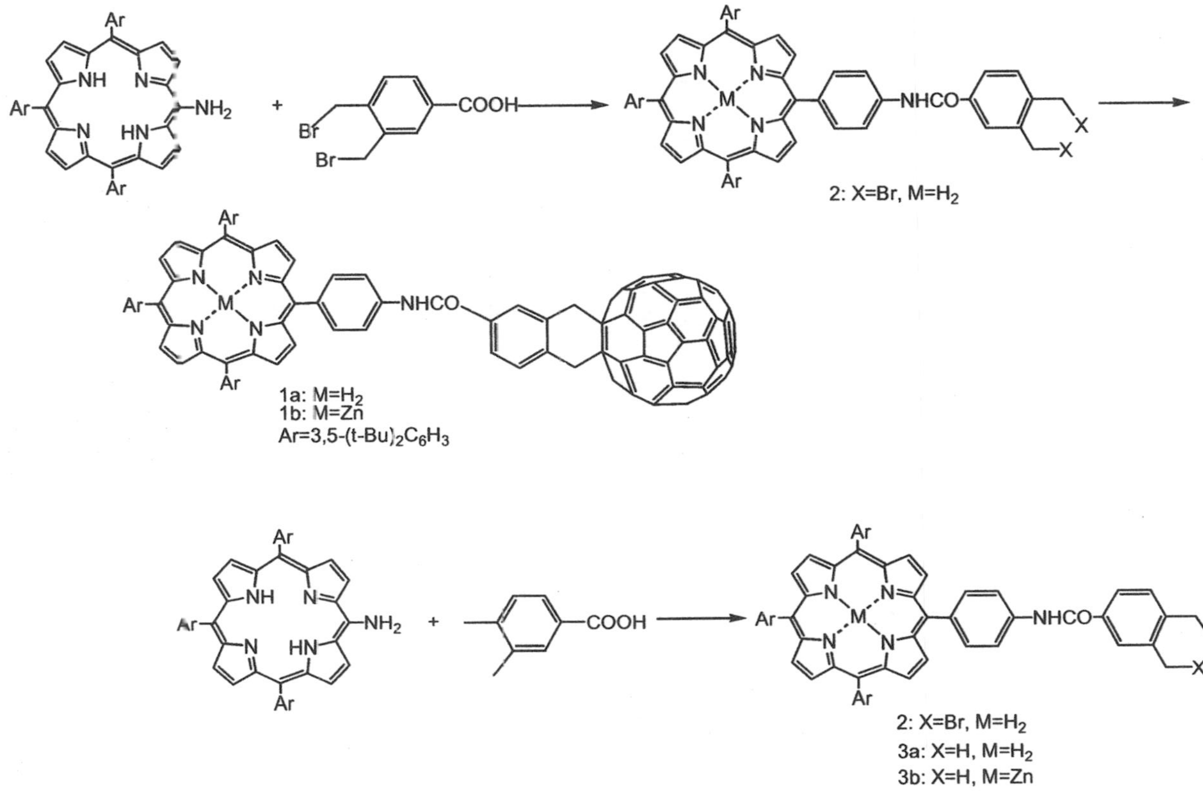

35

Scheme 1-15 Synthesis route of fullerene-linked porphyrin.(Reproduced with permission by H. Imahori et al., *J. Am. Chem. Soc.*, **118** (1996) 11771, ACS)

compounds **3a** and **4** are obtained by the coupling reaction of the aminoporphyrin and 3,4-bis(bromomethyl)benzoic acid and by the Diels-Alder reaction of the corresponding 1,2-bis(bromomethyl) benzene derivative and C_{60}, respectively. Zinc complexes **1b** and **3b** are prepared by treatment of **1a** and **3a** with zinc acetate in chloroform. 270 MHz ^{1}H-NMR spectrum of **3a** in CDCl$_3$-d_3 is shown in Fig. 1.22.[59] The absorption spectra of 3a and fullerene-free zinc porphyrin are shown in Fig. 1.23.[59] There is no evidence for strong interactions between the two moieties in the ground state. The absorption due to the C_{60} is much weaker and broader compared with the porphyrin.

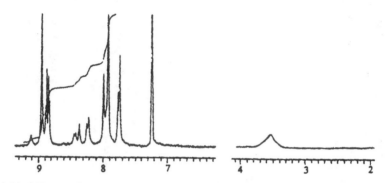

Fig. 1.22 270 MHz ^{1}H NMR spectrum of fullerene-linked porphyrin in CDCl$_3$. (Reproduced with permission by H. Imahori et al., *J. Am. Chem. Soc.*, **118** (1996) 11771, ACS)

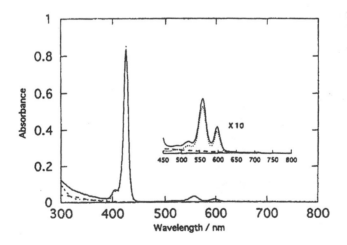

Fig. 1.23 Absorption spectra of fullerene-linked porphyrin and fullerene-free porphyrin. (Reproduced with permission by H. Imahori et al., *J. Am. Chem. Soc.*, **118** (1996) 11771, ACS)

1.3.6 New phthalocyanines

Phthalocyanines show intense absorption in the red region and their appearance in the solid state ranges from dark blue to metallic bronze to green, depending on the central metal, crystalline form and particle size. They are widely used as colorants for plastics, inks, fabrics and automobile paints. In addition, they have also been found to exhibit interesting optical, magnetic, catalytic, semiconductive and photoconductive properties. In this section, syntheses and properties of some interesting phthalocyanines are introduced.

A number of polymeric compounds based on repeating units containing phthalocyanine moieties have been synthesized to establish their photochemical and electrochemical properties. Polymeric phthalocyanines are interesting compounds that exhibit qualities as catalysts and semiconductors. The synthesis of dioxa-dithia macrocycle-bridged with hexakis(alkylthio) substituent phthalocyanine network polymer (Fig. 1.24) is described as an examples.[61] The route of the phthalocyanine network polymer is shown in Scheme 1-16. Tetracyanodibenzo-[1,7-dithia(12 crown-4)] is heated in *n*- pentanol in the presence of 1,8-diazabicyclo[5.4.0] undec-7-ene with stirring and is refluxed for 12 h under argon. After cooling the dark green product is filtered and then washed successively with hot DMF, hot DMSO then dried. The precipitation above is carried out until no more impurity is observed in the filtrate (TLC). It is then dried in vacuum at 100°C. The product is soluble in conc. H_2SO_4. The nickel-, cobalt- and zinc-containing phthalocyanine network polymers arc also synthesized by a modification of the above procedure. Table 1.2 shows the wavelength and absorption coefficients of the UV-vis spectra of phthalocyanine network polymers. Table 1.3 shows the thermal properties of network polymers. The initial decompositions of the network polymers determined by thermogravimetry occur

Table 1.2   Wavelength and absorption coefficients of the UV/VIS spectra of polymeric phthalocyanines

Compound	Solvent	λ (nm) ($10^{-4} \varepsilon$ (1 $mol^{-1}cm^{-1}$))				
6	H_2SO_4	860 (4.52)*,	767(4.40),	523(2.76),	320(7.30),	220(3.06)
7	H_2SO_4	876 (10.20),	800(5.12),	490(3.04),	318(9.38),	236(6.82)
8	H_2SO_4	870 (13.62),	636(6.28),	485(5.82),	332(14.20),	221(9.02)
9	H_2SO_4	880 (6.82),	760(4.06),	506(2.46),	318(7.22),	226(4.10)

(Reproduced with permission by A.G. Gurek, O. Bekaro Glu, *J. Porphyrins Phthalocyanines*, **1** (1997) 67, John Wiley & Sons Limited)

Table 1.3   Thermal properties of polymeric phthalocyanines

Compound	Initial decomposition temp.	Main decomposition temp.
6	380	620
7	420	625
8	375	580
9	430	630

(Reproduced with permission by A.G. Gurek, O. Bekaro Glu, *J. Porphyrins Phthalocyanines*, **1** 1997) 67, John Wiley & Sons Limited)

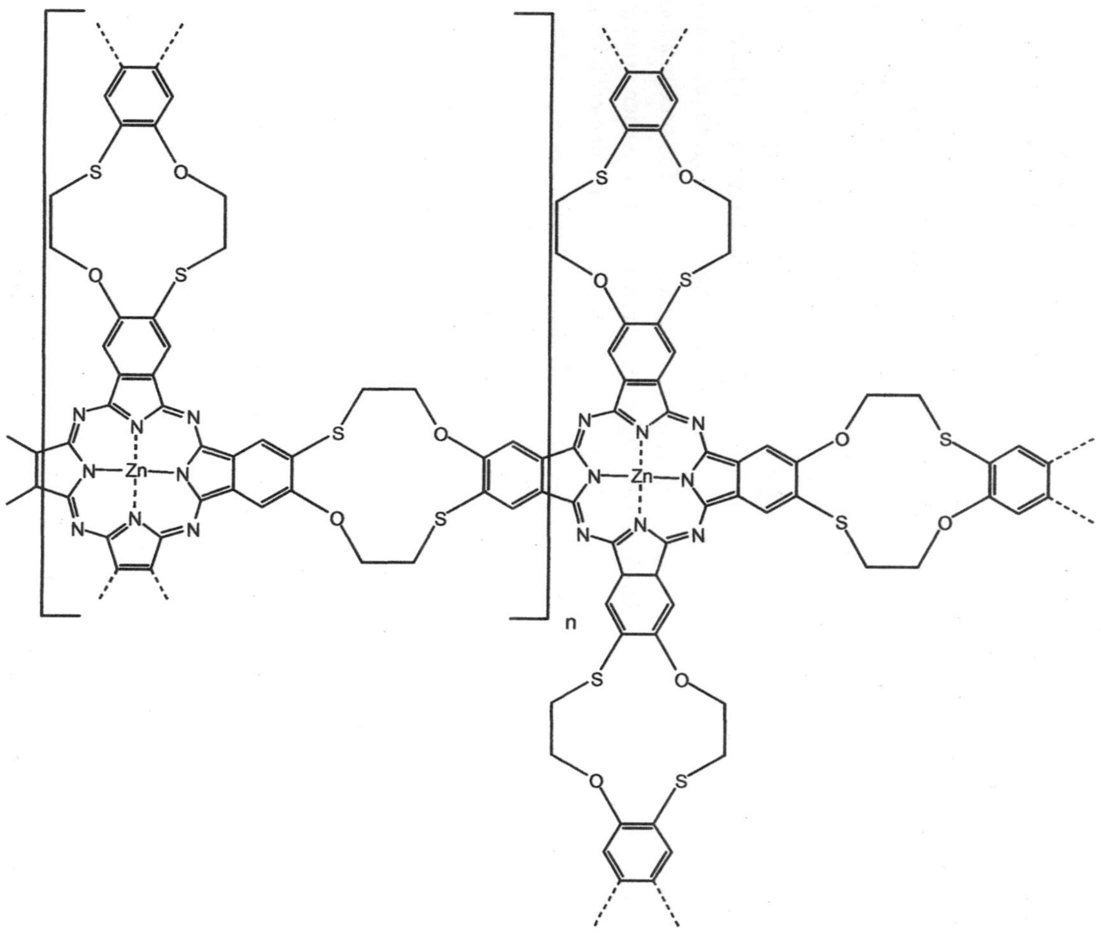

Fig. 1.24 Structure of polymeric phthalocyanines. (Reproduced with permission by V.G. Ramsey, *Biochem. Prep.*, **3** (1953) 39, John Wiley & Sons Limited)

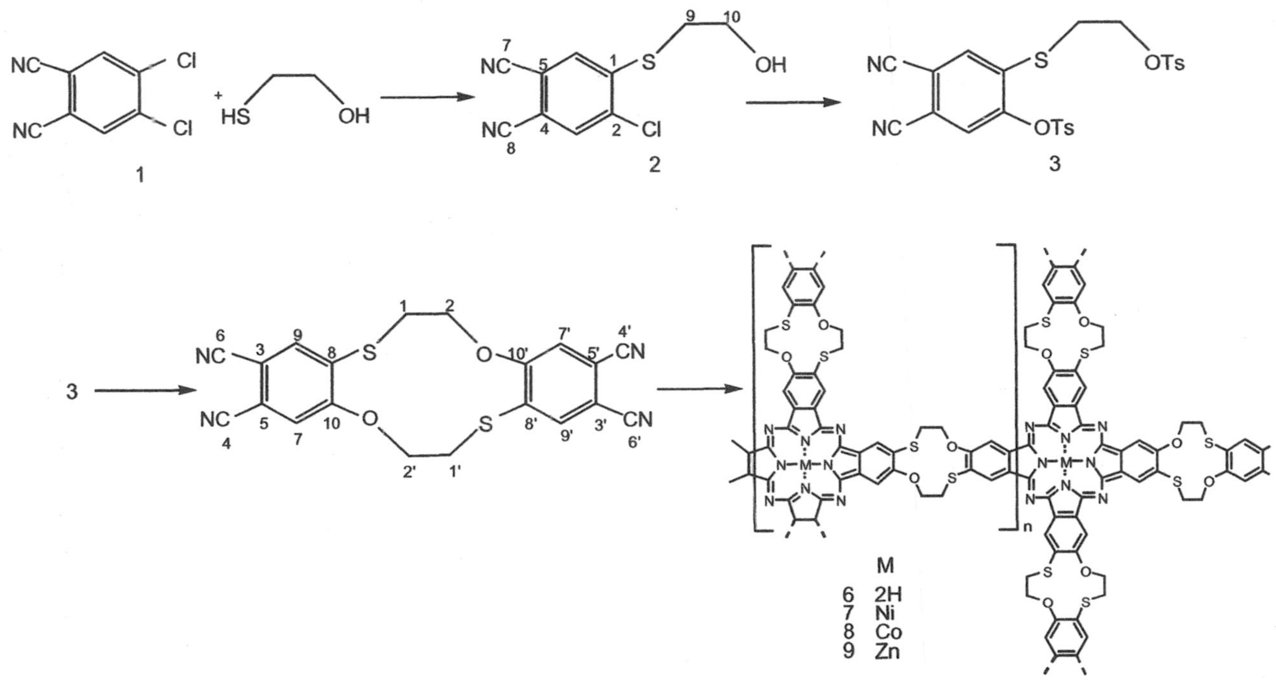

Scheme 1-15 Synthesis route of phthalocyanine network polymer. (Reproduced with permission by A.G. Gurek, O. Bekaro Glu, *J. Porphyrins Phthalocyanines*, 1 (1997) 67, John Wiley & Sons Limited)

above 370°C and show some changes depending on the central metal ions, while the main decomposition points around 580-630°C are in the following order: Co < metal-free < Ni < Zn. The electrical conductivities of the network polymer are given in Table 1.4, measured at room temperature in air and in vacuum in the polycrystalline form as a sandwich between two gold plates. Co- and metal-free network polymers show higher electrical conductivity.

Table1.4 Electrical conductivity of the polymeric and dimeric phthalocyanines

Compound	σ_{dc}(S cm^{-1})	
	In air	In vacuum
Co- network polymer	1.0×10^{-9}	8.50×10^{-10}
Zn- network polymer	4.3×10^{-8}	4.30×10^{-8}
Metal-free network polymer	7.0×10^{-6}	1.35×10^{-7}

(Reproduced with permission by A.G. Gurek, O. Bekaro Glu, *J. Porphyrins Phthalocyanines*, **1** (1997) 67, John Wiley & Sons Limited)

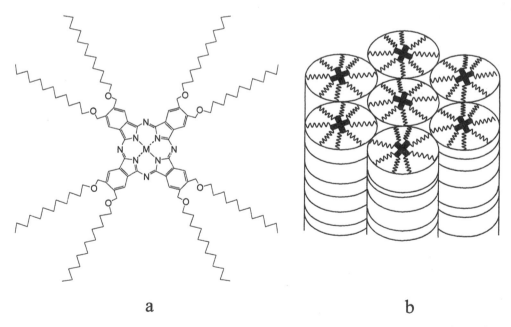

a   b

Fig. 1.25 Liquid crystalline mesophases obtained from octasubstituted metallophthalocyanines. (a) Derivative used. (b) Type of liquid crystals obtained: the phthalocyanines are stacked in columns isolated from each other by molten paraffinic chains. (Reproduced with permission by C. Piechocki, J. Simon, *Nouv. J. Chim.*, **9** (1985) 159)

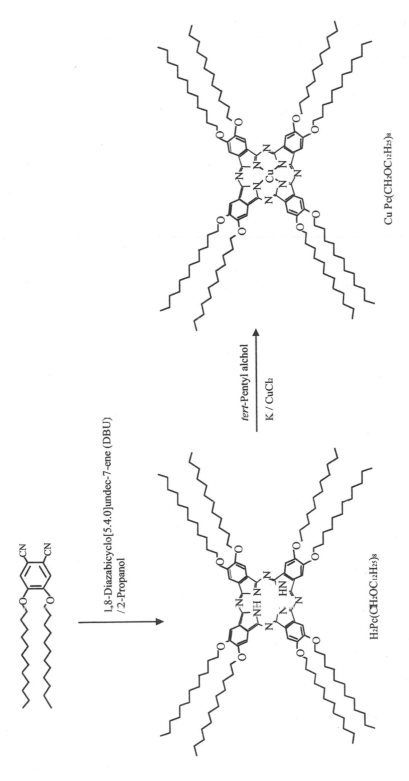

Scheme 1-17 Synthesis route of Cu- phthalocyanines substituted by long paraffinic chains. (Reproduced with permission by C. Piechocki, J. Simon, *Nouv. J. Chim.,* **9** (1985) 159)

Phthalocyanines substituted by long paraffinic chains have recently been described.[61] These derivatives forming liquid crystals in which macrocyclic rings are stacked in columns, the different columns forming a hexagonal array as shown in Fig. 1.25. The synthesis of Cu-phthalocyanines substituted by long paraffinic chains is mentioned as an example. The synthesis route is shown in Scheme 1-17. The phthalocyanines substituted by long paraffinic chains ($Pc(CH_2OC_{12}H_{25})_8$) and copper chloride are heated in *t*-pentyl alcohol at 100°C under nitrogen and refluxed for 2 h. $CuPc(CH_2OC_{12}H_{25})_8$ is purified using alumina gel column chromatography and recrystallized from heptane. The absorption maximum is 680 nm in benzene. The discotic column liquid crystal is formed by phase transition at 66°C.[62]

References

1. V.G. Ramsey, *Biochem. Prep.,* **3** (1953) 39.
2. H. Muir, A Neuberger, *Biochem. J.,* **45** (1949) 163.
3. H. Fischer, H. Orth, *Die Chemie des Pyrrols*, Vol. II, 414, Academische Verlag. Leipzig (1937) (in German).
4. R. Lemberg, J. Parker, *J. Exp. Biol. Med. Sci.,* **30** (1952) 163.
5. H.H. Inhoffen, J.H. Fuhrop, H. Voigt, H. Brockmann Jr, *Ann. Chem.,* **695** (1966) 133.
6. H.W. Whitlosk, R. Hanauer, *J. Org. Chem.,* **33** (1968) 2169.
7. J.B. Paine III, W.B. Krishner, D.W. Moskowitz, D. Dolphin, *J. Org. Chem.,* **41** (1976) 3857.
8. U. Eisner, R.P. Linsted, E.A. Parkes, E. Stephen, *J. Chem. Soc.,* 1655 (**1956**).
9. P. Rothemund, *J. Am. Chem. Soc.,* **57** (1935) 2010.
10. A.D. Adler, F.R. Longo, J. Finarelli, J. Goldmacher, J. Assour, L. Korsakoff, *J. Org. Chem.,* **32** (1967) 476.
11. D.W. Thomas, A.E. Martell, *J. Am. Chem. Soc.,* **81** (1959) 5111.
12. A. Treibs, *Justus Liebigs Ann. Chem.,* **728** (1969) 115.
13. J.B. Buchler, K.L. Lay, *Inorg. Nucl. Chem. Letters,* **10** (1974) 297.
14. J.W. Buchler, G. Eikelman, J. Puppe, K. Rohbock, H.H. Schneehage, D. Weck, *Justus Liebigs Ann. Chem.,* **745** (1971) 135.
15. M. Tsutsumi, M. Ichikawa, F. Vohwinkel, K. Suzuki, *J. Am. Chem. Soc.,* **88** (1966) 854.
16. M. Tsutsumi, R.A. Velapoldi, K. Suzuki, F. Vohwinkel, M. Ichikawa, T. Koyano, *J. Am. Chem. Soc.,* **91** (1969) 6262.
17. A.D. Adler, F.R. Longo, F. Kampas, J. Kim, *J. Inorg. Nucl. Chem.,* **32** (1970) 2443.
18. J.W. Buchler, L. Puppe, K. Rohbock, H.H. Schneehage, *Ann. N. Y. Acad. Sci.,* **206** (1973) 116.
19. H.W. Whitlock, R. Hanauer, M.Y. Oester, B.K. Bower, *J. Am. Chem. Soc.,* **91** (1969) 7485.
20. R.P. Linstead, *J. Chem. Soc.,* (**1934**) 1016.
21. H. Tomoda, S. Saito, S. Ogawa, S. Shiraishi, *Chem. Lett.,* 1277 (**1980**).

22. P.J. Brach, S. J. Grammatica, O.A. Ossanna, L. Weinberger, *J. Heterocycl. Chem.,* **7** (1970) 1403.

23. M.A. Petit, V. Plichon, H. Belkacemi, *New J. Chem.,* **13** (1989) 459.

24. M.A. Petit, T. Thami, S. Sirlin, D. Lelievre, *New J. Chem.,* **15** (1991) 71.

25. K. Uosaki, J. Ueda, *J. Chem. Soc., Perkin Trans. II,* **1331** (1992).

26. N. Fukuda, *Bull. Chem. Soc. Jpn.,* **34** (1961) 884.

27. D. Wohrle, G. Meyer, *Makromol. Chem.,* **181** (1980) 2127.

28. J.F. van der Pol, E. Neeleman, J.C. van Miltenburg, J.W. Zwikker, R.J.M. Nolte, W. Drenth, *Macromolecules,* **23** (1990) 155.

29. C.C. Leznoff, S. Vigh, P.I. Svirskaya, S. Greenberg, D.M. Drew, E. Ben-Hur, I. Rosenthal, *Photochem. Photobiol.,* **49** (1989) 279.

30. *Inorg. Synth.* Vol. XX (1980), p59.

31. A.D. Cian, M. Moussari, J. Fischer, R. Weiss, *Inorg. Chem.,* **24** (1985) 3162.

32. *Inorg. Synth.* Vol. XX (**1980**), p97.

33. J. Yao, H. Yonehara, C. Pac, *Bull. Chem. Soc. Jpn.,* **68** (1995) 1001.

34. M. Handa, A. Suzuki, S. Shoji, K. Kasuga, K. Sogabe, *Inorg. Chim. Acta,* **230** (1995) 41.

35. G.E. Bossard, M.J. Abrams, M.C. Darkes, J.F. Vollano, R.C. Brooks, *Inorg. Chem.,* **34** (1995) 1524.

36. A. Meller, A. Ossko, *Monatsh. Chem.,* **103** (1972) 150.

37. K. Kasuga, S. Nagao, T. Fukumoto, M. Harada, *Polyhedron,* **15** (1995) 69.

38. J. Janczak, R. Kubiak, A. Jezierski, *Inorg. Chem.,* **34** (1995) 3505.

39. Y. Kagaya, H. Isago, *Chem. Lett.,* 1957 (**1994**).

40. C. Franco, C. McLendon, *Inorg. Chem.,* **23** (1984) 2370.

41. Y. Yamamoto, S. Noda, N. Nanai, I. Okura, Y. Inoue, *Bull. Chem. Soc. Jpn.,* **62** (1989) 2152.

42. J. Hirota, T. Takeno, I. Okura, *J. Photochem. Photobiol. A: Chem.,* **77** (1994) 29.

43. Y. Amao, T. Kamachi, I. Okura, *J. Photochem. Photobiol. A: Chem.,* **98** (1996) 59.

44. Y. Amao, T. Hiraishi, I. Okura, *J. Mol. Catal., A: Chem.,* **126** (1997)13.

45. S. Nishitani, N. Kurata, Y. Sakata, S. Misumi, M. Migita, T. Okada, N. Mataga, *Tetrahedron Lett.,* **22** (1981) 2099.

46. S. Nishitani, N. Kurata, Y. Sakata, A. Karen, M. Migita, T. Okada, N. Mataga, *J. Am. Chem. Soc.,* **105** (1983) 7771.

47. J. P. Collman, A. O. Chong, G. B. Jameson, R. T. Oakley, E. Rose, E. R. Schmittou, J. A. Ilvers, *J. Am. Chem. Soc.,* **103** (1981) 516.

48. A. Takahashi, M. Ohkouchi, N. Mataga, T. Okada, H. Yamada, K. Maruyama, *J. Am. Chem. Soc.,* **115** (1993) 12137.

49. D. Gust, T.A. Moore, A.L. Morre, F. Gao, D. Luttrull, J.M. DeGraziano, X.C. Ma, L.R. Makings, S.-J. Lee, T.T. Trier, E. Bittersmann, G.R. Seely, S. Woodward, R.V. Bensasson, M. Rougee, F.C. De Schryver, M. Van der Auweraer, *J. Am. Chem. Soc.,* **113** (1991) 3638.

50. J. S. Lindsey, H. C. Hsu, I. C. Schreiman, *Tetrahedron Lett.,* **27** (1986) 4969.

51. J.W. Arbogast, A.P. Darmanyan, C.S. Foote, Y. Rubin, F.N. Diederich, M.M. Alvarez, SJ. Anz, R.L. Whetten, *J. Phys. Chem.,* **95** (1991) 11.
52. C. Reber, L.Yee, J. McKiernan, J.I. Zink, R.S. Williams, W.M. Tong, D.A.A. Ohlberg, R.L. Whetten, F. Diederich, *J. Phys. Chem.,* **95** (1991) 2127.
53. K. Palewska, J. Sworakowski, H. Chojnacki, E.C. Meister, U.P. Wild, *J. Phys. Chem.,* **97** (1993) 12167.
54. J.L. Anderson, Y.-Z. An, Y. Rubin, C.S. Foote, *J. Am. Chem. Soc.,* **116** (1994) 9763.
55. S.-K. Lin, L.-L.Shiu, K.-M. Chien, T.-Y. Luh, T.-I. Lin *J. Phys. Chem.,* **99** (1995) 105.
56. R.R. Hung, J.J. Grabowski, *J. Phys. Chem.,* **95** (1991) 6073.
57. J.W. Arbogast, A.P. Darmanyan, C.S. Foote, Y. Rubin, F.N. Diederich, M.M. Alvarez, SJ. Anz, R.L. Whetten, *J. Phys. Chem.,* **95** (1991) 11.
58. H. Imahori, K. Hagiwara, T. Akiyama, S. Taniguchi, T. Okada, M. Shirakawa and Y. Sakata, *Chem. Lett.,* (**1995**) 265.
59. H. Imahori, K. Hagiwara, M. Aoki, T. Akiyama, S. Taniguchi, T. Okada, M. Shirakawa, Y. Sakata, *J. Am. Chem. Soc.,* **118** (1996).11771
60. Y. Zeng, L. Biczok, H. Linschitz, *J. Phys. Chem.,* **96** (1992) 5237.
61. A.G. Gurek, O. Bekaro Glu, *J. Porphyrins and Phthalocyanines,* **1** (1997) 67.
62. C. Piechocki, J. Simon, *Nouv. J. Chim.,* **9** (1985) 159.

Chapter 2 Characterization of Porphyrins and Phthalocyanines

2.1 Optical Properties

The absorption spectra of regular-type metalloporphyrins consist of several bands, Q, B, N, L and M bands.[1] The Q band is two visible bands in the 500-600 nm region. The lowest energy band, called the a band, is the electronic origin $Q(0,0)$ of the lowest excited singlet state. The higher energy band, called the b band, is the origin $Q(1,0)$. The b band has a molar extinction coefficient in the narrow range of 10^4 mol^{-1} dm^3 cm^{-1}. The B band, called the Soret band, is an extremely intense band in the region 380-400 nm. This band is the origin $B(0,0)$ of the second excited singlet state and has a high molar extinction coefficient in the range of 10^5 mol^{-1} dm^3 cm^{-1}. The N, L and M bands are the ultraviolet (UV) region.

Figure 2.1 shows the absorption spectra of the free-base porphyrins, zinc (II) and iron (III) octaethylporphyrin (OEP) and tetraphenylporphyrin (TPP) as examples.[2] These spectra broadly represent most of the absorption spectral types encountered. In free-base porphyrin, the presence of two protons leads to a reduction of the overall porphyrin symmetry. A consequence is that the visible absorption spectrum in the Q-band region changes from two bands to four bands. The $Q(0,0)$ splits into $Q_x(0,0)$ and $Q_y(0,0)$, separated by about 30 nm. Hydrogenation of the exo pyrrole double bonds causes dramatic changes in the spectra of the porphyrin ring. Chlorin is characterized by stronger bands in the 650-700 nm region.

The absorption spectra of the regular metallophthalocyanines are composed of two very

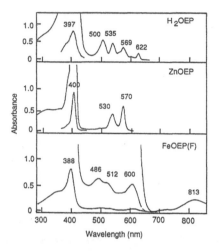
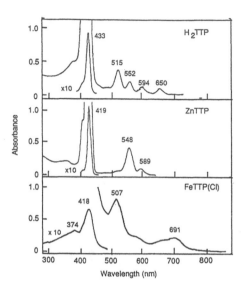

Fig. 2.1 Absorption spectra of OEP and TPP derivatives of metal-free, zinc and iron (III) complex in
dichloromethane. (Reproduced with permission by K. Kalyanasundaram, *Photochemistry
of Polypyridine and Porphyrin Complexes*, p.500, Academic Press, New York (1992))

intense bands in the 300-400 nm region (Soret band) and in the 650-700 nm region (Q band).[3)]
The peak around 700 nm is split into a doublet in the case of metal-free phthalocyanine. This
suggests that metal-free phthalocyanine has a lower symmetry than metallophthalocyanines and
that symmetrically degenerate bands are involved in the transition centered at 700 nm. Most
metallophthalocyanines show three other bands in the UV region. These bands are labeled N, L
and C bands. The N band is the most sensitive to substitutions of the centered metal ion. Typical
absorption spectra of metal-free, zinc and copper phthalocyanines are shown in Fig. 2.2.

2.2   Photophysical Properties of the Photoexcited State

The emission behaviors of metalloporphyrins are classified into the following categories: fluorescent,
phosphorescent, luminescent and radiationless.[4-12)] The Jablonski diagram is shown in Fig. 2.3.
Fluorescent porphyrins (Mg, Zn, Cd, Sn(IV), and Al porphyrins)[4-12)] show moderate fluorescence
with quantum yield in the range 10^{-3}. Lifetimes of the photoexcited singlet state are typically
about 60 ns, largely determined from the integral of the absorption spectra. Phosphorescence is
also observed with quantum yield in the range 10^{-4} to 0.2 at 77 K. Phosphorescence lifetimes are
long, in the range 1-400 ms, determined by the extent of the heavy atom effect. Phosphorescent
porphyrins (Pd, Pt, Ru, Os, and Rh porphyrins)[4-12, 13-54)] show no more than weak prompt

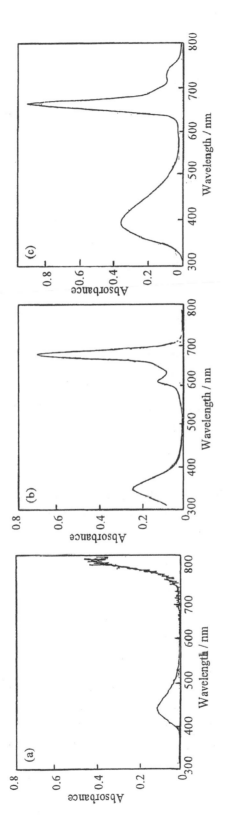

Fig 2.2　Absorption spectra of metal-free(a), zinc(b) and copper(c) phthalocyanines in toluene.

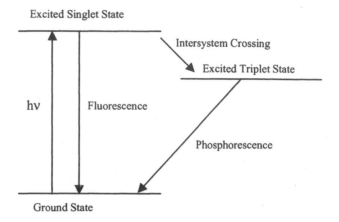

Fig. 2.3 Jablonski diagram (state energy diagram).

fluorescence ($< 10^{-3}$) and phosphorescence yields vary over the entire range from 10^{-4} to 1. Phosphorescence lifetimes are short, typically less than 3.0 ms at room temperature. Luminescent porphyrins (Cr porphyrins) [4-12, 55, 56] contain paramagnetic metals. The emission cannot be characterized as pure fluorescence or phosphorescence. The emission yields are above 10^{-4}, and the emission parameters often show strong temperature dependence. Radiationless porphyrins (Co, Ni, Fe(II), Fe(III), Ag, Sn(II), Pb, Sb, Bi, Mn, Mo, and W porphyrins) [4,5,57,58] show at most weak emission with total yield under 10^{-4}. The lifetime of the excited state is from 6 ps to 60 fs and any prompt emission may be from vibrationally non-relaxed excited states.

Luminescence properties of metallophthalocyanines have been studied extensively.[59-63] The spin multiplicity of the excited state is determined by the nature of the central metal ion and its effect on the magnitude of the spin-orbital coupling. The luminescence properties of typical metallophthalocyanines are shown in Table 2.1. Light closed shell metallophthalocyanines exhibit high fluorescence yields (0.03 to 0.7) and have only weak phosphorescence. [64-67] For the representative metallophthalocyanines, fluorescence lifetimes are on the order of nanoseconds, 6.5 ns for metal-free phthalocyanine, 7.2 ns for magnesium phthalocyanine and 3.8 ns for zinc phthalocyanine.[66, 68-71] Phosphorescence lifetimes are strongly dependent upon the cation species. For lighter metals, magnesium, zinc and cadmium, the lifetime is as long as 0.3 to 1.1 ms at lower temperatures. For heavier metals, the phosphorescence yield increases to the detriment of the fluorescence yield and there is a concomitant decrease in phosphorescence lifetime, indicating that the excited singlet state either reemits a photon to return to the ground state or is interconverted to the triplet state, and no thermal degradation of light energy occurs.

Phosphorescence yields at room temperature are in most cases extremely low; radiationless decay of the lowest excited state back to the ground state by thermal degradation is the preferential major de-excitation pathway for the triplet states. The phosphorescent phthalocyanines (Pd, Pt,

Table 2.1   Luminescence properties of metallophthalocyanines

PcM $M=$	Fluorescence max. (nm)	Phosphorescence max. (nm)	Fluorescence lifetime (ns)	Phosphorescence lifetime (ns)	Φ_F	Φ_P	Φ_T
H_2	699			140	0.7		0.14
Mg	683	1100		1000	0.6	10^{-6}	
Zn	683	1092		1100	0.3	10^{-4}	
Cd	692	1096		350	10^{-2}	10^{-4}	
Cu		1065		3	10^{-4}	10^{-3}	
				0.035	0		0.7
VO	663	1140–1180				10^{-6}	
Pd		990		25	10^{-4}	10^{-3}	
Pt		944		7		10^{-2}	
RhCl	662	987		16	10^{-3}	10^{-3}	
IrCl		957		3.5		10^{-3}	
AlCl			6.8		0.58		0.4
GaCl			3.8		0.31		0.7
InCl			0.37		0.031		0.9

Φ_F: Fluorescence quantum yield; Φ_P: Phosphorescence quantum yield; Φ_T: triplet yield

Ir, and Rh phthalocyanines) [64-67] show no more than weak prompt fluorescence ($< 10^{-3}$) and phosphorescence yields vary over the entire range from 10^{-4} to 10^{-2}. As spin-orbital coupling becomes stronger, both radiative and nonradiative rates increase, but the radiative rates increase faster. This results in higher phosphorescence and total luminescence yields. Nickel and cobalt phthalocyanines, on the other hand, do not show any luminescence. [64-67] This has been attributed the presence of low-lying energy states arising from metal d electrons. In these phthalocyanines the metal states lie below the triplet state of the macrocyclic ligand providing radiationless paths to the ground state. The weak emission is observed in the ultraviolet region. For example, weak emission of magnesium phthalocyanine is observed at 424 nm. Free-base, titanium, zinc, rhodium and copper phthalocyanines also emit in the ultraviolet region.[72-76]

It is almost certain that the photoexcited triplet state of photosensitizers play an important role in the photochemical reaction, for the state has a longer lifetime than that of the photoexcited singlet state. The longer lifetime characteristic of the photoexcited triplet state is a considerable advantage, since at a given concentration of quencher the number of diffusional encounters between a molecule in an excited state and a quencher molecule increases when the lifetime of the excited state is extended. More importantly, recent works have established that separation of photoredox products is more efficient for a triplet process than for the singlet process.[77-81] Thus no electron transfer products are observed when the fluorescent state of chlorophyll is quenched by quinones since reverse electron transfer within the solvent cage is spin allowed.[79-81] In contrast, quenching of the photoexcited triplet state of chlorophyll by duroquinone lead to formation of electron transfer products.[82] In the latter case, spin rephrasing occurs within the solvent cage before the thermodynamically favorable reverse electron transfer takes place. As a result, ions escape from the solvent cage.

As the photoexcited triplet state of a photosensitizer plays an important role as mentioned above, the photophysical properties of the photoexcited triplet state for typical photosensitizers are summarized in Table 2.2.[83-96] All of the compounds listed in Table 2.2 have the absorption

Table 2.2   Typical photophysical properties of several photosensitizers for excited triplet state

Compound	λ_{max} (nm)	τ_T (300K) (μs)	Φ_T	E_T (eV)	Ref
Ru(bpy)$_3$$^{2+}$	452	0.65	1.0	2.1	83
Profravine	445	20		2.1	84
ZnTPPS	555	1500			85
ZnTMPyP	560	655	0.9	1.57	86
ZnTPP	605	1200	0.88	1.59	87, 88
TPP	647	1380	0.82	1.43	89
Chlorophyll-a	661	1000	0.6	1.33	90-92
Pc	698	140	0.14	1.24	93-95
ZnPcTS	690	245	0.56	1.12	96

λ_{max} : The peak of the absorption band with maximum wavelength
τ_T: Lifetime of the excited triplet state
Φ_T: Quantum yield for the formation of the excited triplet state
E_T: The energy of the excited triplet state

band in the visible region, and these compounds are used as photosensitizers. As shown in the table, the lifetimes of porphyrins are longer than that of $Ru(bpy)_3^{2+}$, proflavin or phthalocyanines. With respect to the lifetime of the photoexcited triplet state, porphyrins, especially zinc porphyrin, has advantageous features. Furthermore zinc porphyrins have higher quantum yields, near unity, for the formation of the photoexcited triplet state.

As phthalocyanines absorb light with longer wavelength, these compounds are able to collect up to 50% sunlight.[97] However, these compounds are apt to aggregate in aqueous solution.[98,99] Such dimers possess very short photoexcited lifetimes and are not good photosensitizers. In contrast, porphyrins do not aggregate as easily as phthalocyanines.

A large number of works on photoreduction of methyl viologen with chloroplast as a photosensitizer on irradiation of visible light have been reported.[100-102] As the activity of chloroplast decreases very rapidly under irradiation, the chloroplast is not suitable for use of long duration. On the other hand, porphyrins and $Ru(bpy)_3^{2+}$ are stable against irradiation.

With respect to photophysical properties, porphyrins are suitable photosensitizers. To be good photosensitizers, furthermore, these compounds have high enough energy of the photoexcited state to induce chemical reactions. In the photoreduction of methyl viologen, a photoexcited photosensitizer is oxidized with methyl viologen. In a few cases a photoexcited photosensitizer is reduced with an electron donor then methyl viologen is reduced with the reduced photosensitizer. The driving force of the reaction depends on the redox potential of the photoexcited state of a photosensitizer. Some redox potentials of the photoexcited triplet state are shown in Table 2.3. These values are estimated by the following equations.[103]

$$E(P^+ / P^*) = E(P^+ / P) - E(P / P^*)$$
$$E(P^* / P^-) = E(P / P^-) + E(P / P^*)$$

$E(P^+ / P^*)$ and $E(P^* / P^-)$ are one-electron oxidation potentials and one-electron reduction potentials, respectively. $E(P / P^*)$ is the energy of the photoexcited triplet state.

Table 2.3   Redox potentials of the excited triplet state of typical photosensitizers

Compound	$E(P^+/P^*)$ (V)	$E(P^*/P^-)$ (V)
TPP	-0.20	0.58
ZnTPPS	-0.75	0.45
ZnTPPC	-0.80	0.35
ZnTPP	-0.55	0.48
ZnTMPyP	-0.45	0.78
PdTPPS	-0.60	0.74
PdTMPyP	-0.41	1.18
$Ru(bpy)_3^{2+}$	-0.80	0.80

(Reproduced with permission by K.Kalyanasundaram, M. Nerman-Spallart, *J. Phys. Chem.*, **86** (1982) 613, ACS)

2.3   Photochemical Reactions of Excited States

Many reactions are initiated by light absorption. In this section basic photochemical processes are reviewed. A summary of the processes occurring in the photochemical reaction is as follows.

Primary absorption: $S_0 \rightarrow S_1$ (reaction rate:I) followed by vibration and rotational relaxation

Physical processes:

Fluorescence: $S_1 \longrightarrow S_0 + hv$ $\qquad\qquad$ $(k_f[S_1])$

Intersystem crossing: $S_1 \rightarrow T_1$ $\qquad\qquad$ $(k_{isc}[S_1])$

Phosphorescence: $T_1 \rightarrow S_0 + hv$ $\qquad\qquad$ $(k_p[T_1])$

Internal conversion: $S_1 \rightarrow S_0$ $\qquad\qquad$ $(k_i[S_1])$

Intersystem crossing: $T_1 \rightarrow S_1$ $\qquad\qquad$ $(k'_i[S_1])$

Triplet electronic energy transfer: $T_1 + S_0 \rightarrow S_1 + T_1'$ $\qquad$ $(k_e[T_1][M])$

where S and T denote a singlet state and a triplet state, respectively. M is the reaction substrate. Total reaction rate, I, and intersystem crossing rate, $k_{ISC}[S_1]$, are expressed as follows.

$$I = k_i[S_1] + k_f[S_1] + k_{ISC}[S_1] \qquad\qquad (2\text{-}1)$$

$$k_{ISC}[S_1] = k_p[T_1] + k_i' [T_1] + k_e[T_1][A] \qquad\qquad (2\text{-}2)$$

$$[S_1] = I / (k_i + k_f + k_{ISC}) \qquad\qquad (2\text{-}3)$$

$$[T_1] = k_{ISC}[S_1] / (k_p + k_i' + k_e[M]) \qquad\qquad (2\text{-}4)$$

Quantum yield (usually indicated by the symbol Φ) of phosphorescence in the presence of (Φ_p) and in the absence of reaction substrate M (Φ_p^0) is expressed as follows.

$$\Phi_p = k_p[T_1] / I \qquad\qquad (2\text{-}5)$$

$$\Phi_p^0 = k_p[T_1]' / I \qquad\qquad (2\text{-}6)$$

Where $[T_1]'$ is the concentration of photoexcited triplet state in the absence of reaction substrate M.

$$\Phi_p^0 / \Phi_p = [T_1]' / [T_1] = (k_p + k_i' + k_e[M]) / (k_p + k_i') \qquad\qquad (2\text{-}7)$$

$$\Phi_p^0 / \Phi_p = 1 + k_e /(k_p + k_i') [M] \qquad\qquad (2\text{-}8)$$

$$I_p^0 / I_p = 1 + k_e /(k_p + k_i') [M] \qquad\qquad (2\text{-}9)$$

where I_p^0 and I_p are phosphorescence intensities in the presence of and in the absence of reaction substrate M, respectively. Eqs. 2-8 and 2-9 are the Stern-Volmer equation.

Typical photochemical reactions are classified into the following categories: (1) photoionization and (2) chemical processes. Typical ionization processes are penning, dissociative, collisional and associative as follows:

Penning ionization: $A^* + B \longrightarrow A + B^+ + e^-$

Dissociative ionization: $A^* + B\text{-}C \longrightarrow A\text{-}B^+ + C + e^-$

Collisional ionization: $A^* + B \longrightarrow A^+ + B^- + e^-$ or $A^+ + B + e^-$

Associative ionization: $A^* + B \longrightarrow AB^+ + e^-$

On the other hand, typical chemical processes are dissociation, addition, insertion, abstraction, fragmentation, isomerization and dissociative excitation.

Dissociation: $A\text{-}B^* \longrightarrow A + B$

Addition or insertion: $A^* + B \longrightarrow AB$

Abstraction or fragmentation: $A^* + B \longrightarrow C + D$

Isomerization: $A \longrightarrow A'$

Dissociative excitation: $A^* + C\text{-}D \longrightarrow A + C^* + D$

Quantum yield or efficiency of photoreaction, i.e., the ratio of the number of molecules reacting to the number of photons absorbed, is a very important value.

Φ = (Number of produced molecules by photochemical reaction) / (Number of photons absorbed)

For example, Φ of photoinduced electron transfer processes via the photoexcited singlet state of chlorophyll in the reaction center is estimated to be 1.0 in the photosynthesis reaction.

2.4   Techniques for the Study of Excited States and Photoreactions

The photophysical and photochemical properties of the excited state of porphyrins and phthalocyanines are often studied using steady state luminescence (fluorescence and phosphorescence) spectrum, luminescence lifetime measurement and time-resolved spectroscopy using laser flash photolysis. In this section, some techniques for the investigation of the photoexcited states and photoreactions are described.

Steady state luminescence spectrum is widely used for the investigation of energy transfer, electron transfer and photoreaction in the excited states. The setup for luminescence spectrum measurement is shown in Fig. 2.4. A xenon arc lamp is usually used as the excitation light source. The luminescence is detected using a photomultiplier. In this system two monochromators are required for luminescence detection (M_1) and excitation source (M_2). For luminescence spectrum detection, M_1 is scanned with fixation of M_2. For excitation spectrum measurement, M_2 is scanned with fixation of M_1. As an example, the fluorescence spectra of water soluble zinc porphyrin (ZnTMPyP) and viologen-linked zinc porphyrin are shown in Fig. 2.5.[104] Although the shapes of these spectra are almost the same, the intensity of viologen-linked zinc porphyrin is lower than that of ZnTMPyP. This shows that the fluorescence of zinc porphyrin is quenched by the bonded viologen. The energy transfer, electron transfer and photoreaction in the excited states are studied using the luminescence spectrum.

Luminescence lifetime is widely used for studies on the processes of energy transfer, electron transfer and photoreaction in the excited states. Luminescence lifetime is measured using time-correlation single photon counting. The setup for luminescence lifetime measurement is shown in Fig. 2.6. Argon ion or Nd-YAG laser is usually used as the pulse laser source. For photon counting only a few molecules should be excited. Thus, the intensity of the pulse lamp is chosen to be so low that only one molecule is emitted per excitation pulse. This single photon must be counted. An electronic discriminator guarantees that only one photon is measured at a time. Other photons are ignored after the first one has reached the detector. The photons emitted are detected by a photomultiplier and electronically counted into a multi-channel analyzer. Further, the time at which time the photon is statistically emitted after excitation must be determined. To obtain the correlation with the moment of emission, a time window is electronically moved along the time axis. The related time can be found via the electronic lamp by which the time window has been advanced. It is evident that statistically more photons are emitted at short time intervals

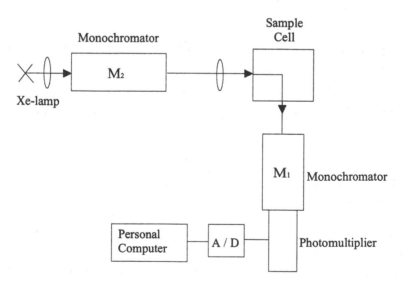

Fig. 2.4 The setup for luminescence spectrum measurement.

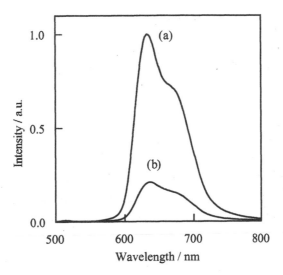

Fig. 2.5   Fluorescence spectra of (a) $ZnP(C_3)_4$ and (b) $ZnP(C_3V)_4$ in water. The excitation wavelength is 438 nm. (Reproduced with permission from *J. Porphyrins Phthalocyanines,* **2** (1998) 201, John Wiley & Sons Limited)

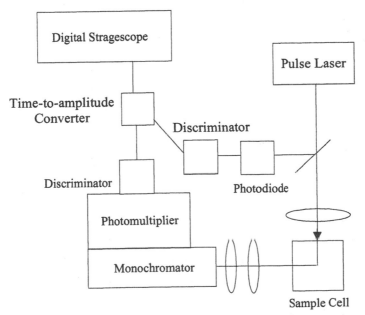

Fig. 2.6 The setup for luminescence lifetime measurement.

after the pulse is fired. Consequently, in the multichannel analyzer the related channels are more filled than the later ones. By firing more than 10,000 pulses on the sample a relatively good signal-to-noise ratio is founded. The problem is that the intensity of the excitation pulse also hits the detector so that its signal overlaps with the luminescence intensity measured by the multichannel analyzer. For this reason, in the first step of evaluation deconvolution methods must be used to separate the pulse intensity with its tailing from the beginning of the exponential decay of the luminescence. In the second step, the measured data must be set by exponential curves. The type of necessary exponential fit (one or more exponential functions) allows one to decide between one or more degradation pathways. Lifetimes are estimated from the exponents in the evaluation formula. As an example, the fluorescence decay of water-soluble viologen-linked zinc porphyrin is shown in Fig. 2.7.[104] The decay consists of two components and the fluorescence lifetimes are

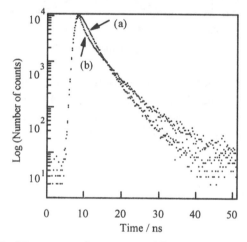

Fig. 2.7   Fluorescence decay curve of $ZnP(C_3V)_4$ in water. The abosrbance at the excitation wavelength (350 nm) is kept to 0.4 for all the sample solution. The solid line indicates the lamp decay. (Reproduced with permission from *J. Porphyrins Phthalocyanines*, **2** (1998) 201, John Wiley & Sons Limited)

0.24 ns and 4.4 ns. The shorter lifetime (τ_s) is the main component and its ratio is more than 80%. The longer lifetime (τ_l) component is a large value compared with that of viologen-free zinc porphyrin by connection of viologen to zinc porphyrin. There may be two conformers: a complexed conformer (the viologen is close to the porphyrin moiety) and extended conformers (the viologen is extended from the porphyrin moiety). The shorter lifetime and longer lifetime components are assigned to the complexed and extended conformers, respectively.[105] From the fluorescence lifetimes, intramolecular electron transfer rate constants (k_{et}) are estimated by the following equation:

$$k_{et} = 1 / \tau_s - 1 / \tau_l.$$

The k_{et} is estimated to be 3.9×10^9 s^{-1}. The electron transfer in the excited states are studied using luminescence lifetime measurement.

Time-resolved spectroscopy using laser flash photolysis is widely used for the investigation of the kinetic and mechanism of photoreactions. The setup for the laser flash photolysis is shown in Fig. 2.8. The nitrogen laser, eximer laser and Nd-YAG laser are usually used as the excitation light source. A xenon arc lamp is used as the monitoring light source. The light beam, after passing through the sample cell, is collimated into the entrance slit of a monochromator. The output signal from a photomultiplier attached to the slit of the monochromator is displayed on a digitizing oscilloscope and the signals are averaged over 128 flashes. The decay to fit this measured data is set by exponential curves. The type of necessary exponential fit (one or more exponential functions) allows one to decide between one or more degradation pathways. Lifetimes are estimated from the exponents in the evaluation formula. In general, the photoreactions are studied using a laser with nanosecond pulse width. The reaction via the photoexcited singlet state (intramolecular photoinduced electron transfer processes, photoreaction in the photosynthesis reaction center and others) are studied using ultra-fast laser flash photolysis with picosecond or femtosecond laser pulse width. As an example, studies on quenching of the photoexcited triplet state of water-soluble zinc porphyrin (ZnTPPS) by methyl viologen using laser flash photolysis are described.[106, 107] Fig. 2.9 shows a typical decay of the photoexcited triplet state of ZnTPPS. The decay consists of a single lifetime component without methyl viologen and the lifetime is estimated to be ca. 650 μs (Fig. 2.9a). In the presence of methyl viologen, on the other hand, the decay rate increases less

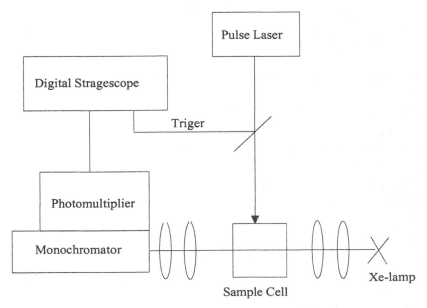

Fig. 2.8 The setup for the laser flash photolysis.

than 10 μs, as shown in Fig. 2.9b, indicating that the photoexcited triplet state of ZnTPPS is quenched by methyl viologen. The quenching rate constant k_q is estimated to be $10^9 \text{mol}^{-1} \text{dm}^3 \text{s}^{-1}$. The quenching efficiency, quenching rate constant and other values in photoexcited states are measured using laser flash photolysis.

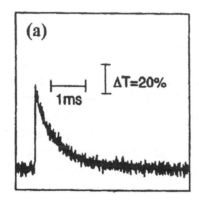 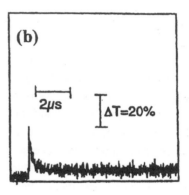

Fig. 2.9   Typical decay of the photoexcited triplet state of ZnTPPS in the absence of (a) and presence of methyl viologen (b). The excitation and monitoring wavelength are 532 and 470 nm. (Reproduced with permission from *J. Mol. Catal. A: Chem.*, 105 (1996) 125, Elsevier Science)

References

1.   M. Gouterman, *The Porphyrins* ed. D. Dolphin, Vol. III, P.1, Academic Press, New York, 1978.
2.   J.W. Bucher in *Porphyrins and Metalloporphyrins* ed. K.M. Smith, p 389, Elsevier, Amsterdam, 1976.
3.   J.A. Shelnutt, *J. Phys. Chem.,* **88** (1984) 6121.
4.   K. Kalyanasundaram, *Phtochemistry of Polypyridine and porphyrincomplexes,* p500, Academic Press, New York, 1992.
5.   M. Gouterman, *J. Chem. Phys.,* **30** (1959)1139.
6.   M. Zerner and M. Gouterman, *Theor. Chim. Acta,* **4** (1963) 44.
7.   A. Antipas, J.W. Buchler, M. Gouterman and P.D. Smith, *J. Am. Chem. Soc.,* **100** (1978) 3015.
8.   P.J. Spellane, M. Gouterman, A. Antipas, S. Kim and Y.C. Liu, *Inorg. Chem.,* **19** (1981) 386.
9.   A.Antipas and M. Gouterman, *J. Am. Chem. Soc.,* **105** (1983) 4896.
10. J.A. Shelnutt, *J. Phys. Chem.,* **88** (1984) 4988.
11. J.A. Shelnutt, *J. Phys. Chem.,* **89** (1985) 4733.
12. H. Sekino and H. Kobayashi, *J. Chem. Phys.,* **86** (1987) 5045.
13. D. Eastwood and M. Gouterman, *J. Mol. Spectrosc.,* **35** (1970) 359.

14. J.B. Callis, M. Gouterman, Y.M. Jones and B.H. Henderson, *J. Mol. Spectrosc.,* **29** (1971) 410.
15. X. Yan, C. Kirmaier and D. Holten, *Inorg. Chem.,* **25** (1986) 4774.
16. D.H. Kim, D. Holten, M. Gouterman and J.W. Buchler, *J. Am. Chem. Soc.,* **106** (1984) 4015.
17. C. Ponterini, N. Serpone, M.A. Bergkamp and T.L. Netzel, *J. Am. Chem. Soc.,* **105** (1983) 4639.
18. T. Kobayashi, K.D. Straub and P.M. Rentzepis, *Photochem. Photobiol.,* **29** (1979) 925.
19. D. Magde, M.W. Windsor, D. Holten and M. Gouterman, *Chem. Phys. Lett.,* **29** (1974) 183.
20. J.B. Callis, M. Gouterman, Y.M. Jones and B.H. Henderson, *J. Mol. Spectrosc.,* **39** (1971) 410.
21. J.B. Callis, J.M. Knowles and M. Gouterman, *J. Phys. Chem.,* **77** (1973) 154.
22. V.V. Sapunov and M.P. Tsvirko, *Opt. Spectrosc. (USSR),* **49** (1980) 152.
23. M.P. Tsvirko, V.V. Sapunov and K.N. Solovev, *Opt. Spectrosc. (USSR),* **34** (1973) 635.
24. L.K. Hanson, M. Gouterman and J.C. Hanson, *J. Am. Chem. Soc.,* **95** (1973) 4822.
25. A.B.P. Lever, B.S. Ramaswamy and S. Licoccia, *J. Photochem.,* **19** (1982) 173.
26. W.A. Lee, K. Kalyanasundaram and M. Gratzel, *Chem. Phys. Lett.,* **107** (1984) 308.
27. R.H. Schmehl and D.G. Whitten, *J. Phys. Chem.,* **85** (1981) 3473.
28. M. Gardiner and A.J. Thomson, *J. Chem. Soc., Dalton Trans.,* (1974) 820.
29. A. Vogler, R. Hisrshmann, H. Otto and H. Kunkely, *Ber. Bunsenges. Phys. Chem.,* **80** (1976) 420.
30. M. Hoshino and K. Yasufuku, *Chem. Phys. Lett.,* **117** (1985) 259.
31. M. Hoshino, H. Seki, K. Yasufuku and H. Shizuka, *J. Phys. Chem.,* **90** (1986) 5149.
32. F.R. Hopf, T.P. O' Brien, W.R. Scheidt and D.G. Whitten, *J. Am. Chem. Soc.,* **97** (1975) 277.
33. G.W. Sovocool, F.R. Hopf and D.G. Whitten, *J. Am. Chem. Soc.,* **97** (1975) 4350.
34. R.C. Young, J.K. Nagle, L.F. Barringer, T.J. Meyer and D.G. Whitten, *J. Am. Chem. Soc.,* **100** (1978) 4773.
35. D.P. Rillema, J.K. Nagle, L.F. Barringer and T.J. Meyer, *J. Am. Chem. Soc.,* **103** (1981) 56.
36. A. Vogler and H. Kunlely, *Ber. Bunsenges. Phys. Chem.,* **80** (1976) 425.
37. B.M. Hoffman and Q.H. Gibson, *Proc. Nat. Acad. Sci. USA,* **75** (1978) 21.
38. A. Antipas, J.W. Buchler, M. Gouterman and P.D. Smith, *J. Am. Chem. Soc.,* **100** (1978) 3015.
39. D.P. Rillema, J.K. Nagle, L.F. Barringer and T.J. Meyer, *J. Am. Chem. Soc.,* **103** (1982) 56.
40. I. Okura and N. Kim-Thuan, *J. Chem. Soc., Chem. Commun.,* (1981) 84.
41. M. Barley, D. Dolphin, B.R. James, C. Kirmaier and D. Holten, *J. Am. Chem. Soc.,* **106** (1984) 3937.
42. C.D. Tait, D. Holten, M. Barley, D. Dolphin and B.R. James, *J. Am. Chem. Soc.,* **107** (1985) 1930.
43. J. Rodriguez, L. McDowell and D. Holten, *Chem. Phys. Lett.,* **147** (1988) 235.
44. L.M.A. Levine and D. Holten, *J. Phys. Chem.,* **92** (1988) 714.

45. D. Kim, Y.O. Su and T.G. Spiro, *Inorg. Chem.*, **25** (1986) 3993.

46. B.A. Crawford and M.R. Ondrias, *J. Phys. Chem.*, **93** (1989) 5055.

47. T. Kitagawa, M. Abe and H. Ogishi, *J. Chem. Phys.*, **69** (1978) 4516.

48. M. Abe, T. Kitagawa and H. Ogishi, *J. Chem. Phys.*, **69** (1978) 4526.

49. T.G. Sprio, *Iron Porphyrins* ed. A.B.P. Lever and H.B. Gray, vol. 2, p. 89 Addison-Wesley, Reading, Mass., 1983.

50. N. Serpone, T.L. Netzel and M. Gouterman, *J. Am. Chem. Soc.*, **104** (1982) 246.

51. G. Ponterini, N. Serpone, M.A. Bergkamp and T.L. Netzel, *J. Am. Chem. Soc.*, **105** (1983) 4639.

52. A. Vogler, J. Kisslinger and J.W. Buchler, *Optical Properties and Structure of Tetrapyrroles* ed. G. Blauer and H. Sund, p 107, de Gruyter, Berlin, 1985.

53. R. Greenhorn, M.A. Jamieson and N. Serpone, *J. Photochem.*, **27** (1984) 287.

54. N. Serpone, M.A. Jamieson and T.L. Netzel, *J. Photochem.*, **15** (1981) 295.

55. M. Gouterman, L.K. Hanson, G.E. Khalil, W.R. Leenstra and J.W. Buchler, *J. Chem. Phys.*, **62** (1975) 2343.

56. B.M. Dzhagarov, P.N. Dylko, K.I. Salokhiddinov and G.P. Gurinovich, *Proceedings of 3rd All-Union Conference on Complex Organic Compound Lasers and their Applications, Uzhgorod, 1980* p111, Institute of Physics, Minsk.

57. V.W. Chirvonyi, B.M. Dzhagarov, Y.V. Timinskii and G.P. Gurinovich, *Chem. Phys. Lett.*, **70** (1983) 79.

58. T. Kobayashi, D. Huppert, K.D. Strab and P.M. Rentzepis, *J. Chem. Phys.*, **70** (1979) 1720.

59. D. Dmietrierskii, V.L. Ermolaev and A.N. Terenin, *Dokl. Acad. Nauk. SSSR*, **114** (1957) 751.

60. V.B. Evstigneev, A.A. Krasnoskii and V.A. Gavrilova, *Dokl. Acad. Nauk. SSSR*, **70** (1950) 261.

61. V.F. Grachkovskii, *Dokl. Acad. Nauk. SSSR*, **82** (1953) 739.

62. F.F. Litvin and R.I. Personov, *Fiz. Prob. Spektropii*, **1** (1963) 229.

63. G.N. Lyalin and G.I. Kobyshev, *Opt. i. Spectroskopiya*, **15** (1963) 253.

64. E.R. Menze, K.E. Rieckhoff and E-M. Vogit, *J. Chem. Phys.*, **58** (1973) 5726.

65. P.S. Vincett, E-M. Vogit and K.E. Rieckhoff, *J. Chem. Phys.*, **55** (1971) 4131.

66. J.H. Brannon and D. Magde, *J. Am. Chem. Soc.*, **102** (1980) 62.

67. J. McVie, R.S. Sinclair and T.G. Truscott, *J. Chem. Soc., Faraday Trans. II*, **1870** (1978).

68. G.A. Williams, B.N. Figgis, R. Mason, S.A. Mason and P.E. Fielding, *J. Chem. Soc., Dalton Trans.* (1980) 1688.

69. A.N. Sevchenko, *Izv. Akad. Nauk. SSSR Ser. Fiz.*, **26** (1962) 53.

70. W.F. Kosonocky and S.E. Harrison, *J. Appl. Phys.*, **37** (1966) 4789.

71. A.T. Gradyushko, A.N. Sevchenko, K.N. Solovyov and M.P. Tsvirko, *Photochem. Photobiol.*, **11** (1970) 387.

72. Y. Kaneko, T. Arai, H. Sakuragi, K. Tokumaru, and C.Pac, *J. Photochem. Photobiol. A: Chem.*, **97** (1996) 155.

73. Y. Kaneko, Y. Nishimura, N. Takane, T. Arai, H. Sakuragi, N. Kobayashi, D. Matsunaga, C. Pac and K. Tokumaru, *Asian Photochemistry Conference,* Hong Kong (1996) p. 135.

74. Y. Kaneko, Y. Nishimura, T. Arai, H. Sakuragi, K. Tokumaru and D. Matsunaga, *J. Photochem. Photobiol. A: Chem.,* **89** (1995) 37.

75. G. Frraudi and S. Muralidharan, *Inorg. Chem.,* **22** (1983) 1369.

76. N. Kobayashi and A.B.P. Lever, *J. Am. Chem. Soc.,* **109** (1987) 7433.

77. A.K. Chibisov, *Russ. Chem. Rev.,* **50** (1981) 1169.

78. J.R. Darwent, I. McCubbin and D. Phillips, *J. Chem. Soc., Faraday Trans. 2,* **78** (1982) 347.

79. A.A. Lamola, M.L. Manion, H.D. Roth and G. Tollin, *Proc. Natl. Acad. Sci. USA,* **72** (1975) 3265.

80. P.M. Rentzepis, D. Huppet and G. Tollin, *Biochem. Biophys. Acta,* **440** (1976) 356.

81. D. Holten, M.W. Windsor, W.W. Pearson and M. Gouterman, *Photochem. Phtobiol.,* **28** (1978) 951.

82. J.M. Kelly and G. Porter, *Proc. Roy. Soc. Lond. A,* **319** (1970) 319.

83. V. Balzani, F. Bolletta, M.T. Gandolpi and M. Maestri, *Topics in Curr. Chem.,* **75** (1978) 1.

84. K. Kalyanasundaram and D. Dung, *J. Phys. Chem.,* **84** (1980) 2551.

85. M.C. Richoux and A. Harriman, *J. Chem. Soc., Faraday Trans, 1,* **78** (1982) 1873.

86. A. Harriman, G. Poter and M.C. Richoux, *J. Chem. Soc., Faraday Trans, 2,* **77** (1981) 1981.

87. A. Harriman, *J. Chem. Soc., Faraday Trans, 2,* **77** (1981) 1281.

88. S.P. Gurinovich and B.M. Dzhagarov, *Izv. Akad. Nauk. SSSR,* **37** (1973) 383.

89. A. Harriman, *J. Chem. Soc., Faraday Trans, 1,* **76** (1980) 1978.

90. K. Kalyanasundaram and G. Porter, *Proc. Roy. Soc. Lond. A,* **364** (1978) 29.

91. G. Seely, *Photochem. Photobiol.,* **27** (1978) 639.

92. P.G. Bowers and G. Porter, *Proc. Roy. Soc. Lond. A,* **296** (1967) 435.

93. M.J. Whalley, *J. Chem. Soc.,* (1961) 866.

94. P.S. Vincett, E.M. Voight and K.E. Rieckhoff, *J. Chem. Phys.,* **55** (1971) 4131.

95. J.G. Villar and L. Lindquist, *C.R. Acad. Sci. Paris, Ser. B,* **264**(1967) 1807.

96. A. Harriman, G. Poter and M.C. Richoux, *J. Chem. Soc., Faraday Trans, 2,* **76** (1980) 1618.

97. M.C. Richoux, *Photogeneration of Hydrogen* ed. A. Harriman and M.A. West, Academic Press, New York 1982.

98. A. Harriman and M.C. Richoux, *J. Photochem.,* **14** (1980) 253.

99. J.R. Darwent, *J. Chem. Soc., Chem. Commun.,* (**1980**) 805.

100. A. Ben-Anotz and M. Gibbs, *Biochem. Biophys. Res. Commun.,* **64** (1975) 355.

101. A.A. Krasnovskii, V.V. Nikandrov, G.P. Brin, I.N. Gogotov and V.P. Oshehepkov, *Dokl. Acad. Nauk. SSSR,* **225** (1975) 771.

102. D. Hoffman, R. Thauer and A. Trabst, *Z. Naturforsch,* **32-C** (1977) 257.

103. K. Kalyanasundaram and M. Nerman-Spallart, *J. Phys. Chem.,* **86** (1982) 6163.

104. Y. Amao, T. Kamachi and I. Okura, *J. Porphyrins Phthalocyanines,* **2** (1998) 201.

105. D. Gust, T.A. Moore, P.A. Liddell, G.A. Nemeth, L.R. Makings, A.L. Moore, D. Barrett, P.J. Pessiki, R.V. Bensasson, M. Rougee, C. Chachaty, F.C. De Scchryver, M. Van der Auweraer,

A.R. Holzwarth and J. S. Connolly, *J. Am. Chem. Soc.,* **109** (1987) 846.

106. Y. Amao and I. Okura, *J. Mol. Catal. A: Chem.,* **103** (1995) 69.

107. Y. Amao and I. Okura, *J. Mol. Catal. A: Chem.,* **105** (1996) 125.

Part II
Application of Porphyrins and Phthalocyanines

Chapter 3   Photoinduced Hydrogen Evolution

3.1   General Introduction to Photoinduced Hydrogen Evolution

Since the second oil crisis in Japan studies on new energy resources to replace fossil fuels have been carried out. Candidates for new energy resources include wind forces, water power, solar energy and others. Among these, solar energy is the most attractive as a future energy resource for the following reasons. Solar energy is huge, virtually limitless and the amount of total energy reaching earth is *ca.* 3.0 $\times 10^{21}$ kJ per year. As solar energy is strongly influenced by weather and seasons, it is necessary to convert solar energy to another form of energy for storage. There are two fundamental ways to convert solar energy into other forms of energy, i.e., thermal conversion and quantum conversion.

In the case of thermal conversion, photons with longer wavelength play an important role but do not have the ability to excite a molecule to higher energy levels. On the other hand, photons with shorter wavelength are responsible for exciting a molecule to initiate the following photochemical reaction in quantum conversion.

In the storage of solar energy by photochemical reactions, the following conditions are required: 1) the reaction must be endothermic, i.e., $\Delta G > 0$; 2) the photoproducts must be stable to store, suitable for transportation and convenient to use; 3) the photoreaction must have high efficiency and high quantum yield.

The photochemical cleavage of water into hydrogen is suitable for the storage of solar energy, because this reaction satisfies the above requirements and hydrogen is expected to be a future fuel resource. Many studies on photoinduced cleavage of water have been carried out recently.[1,2] In this section, the photoinduced cleavage of water and photoinduced hydrogen evolution are described.

3.2 Photoinduced Cleavage of Water

Photosynthesis of green plants and bacterial photosynthesis are typical examples of conversion of solar energy into chemical energy. In photosynthesis, nicotineamide adenine dinucleotide phosphate (NADP) is reduced to NADPH as shown in Scheme 3-1. In this reaction, chlorophyll and water act as a photosensitizer and an electron donor, respectively. This is an effective conversion system of solar energy to chemical energy: the quantum yield is estimated to be *ca.* 1.0.

Many attempts to establish an artificial system mimicking the photosynthesis have been carried

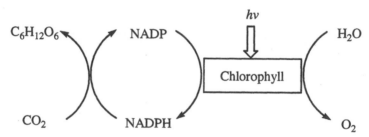

Scheme 3-1 Photosynthesis system.

out. Scheme 3-2 shows typical artificial photosynthesis systems using a photosensitizer, an electron carrier and a catalyst to decompose water to obtain oxygen and hydrogen. An outline of the photoinduced cleavage of water into hydrogen and oxygen and some problems regarding the system are given below.

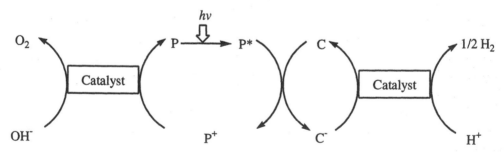

Scheme 3-2 Artificial photosynthesis system. P: Photosensitizer, C: Electron carrier.

Grätzel et al. reported the photoinduced cleavage of water system as shown in Scheme 3-3.[3] In this system, tris(bipyridyl)ruthenium (Ru (bpy)$_3$$^{2+}$) and methyl viologen (MV^{2+}) act as a photosensitizer and an electron carrier, respectively. Since this report, attempts at photoinduced cleavage of water

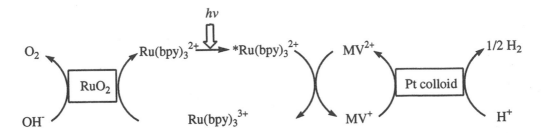

Scheme 3-3 Photoinduced cleavage of water using RuO_2 and platinum colloid. (Reproduced with permission by K. Kalyanasundaram, J. Kiwi, G. Gratzel, *Helv. Chim. Acta*, **61** (1978) 2720)

Table 3.1   Typical photoinduced cleavage of water systems

Photosensitizer	Electron carrier	Catalyst for oxygen evolution	Catalyst for hydrogen evolution	Ref.
$Ru\,(bpy)_3^{2+}$	MV^{2+}	RuO_2	Pt colloid	3
$Ru\,(bpy)_3^{2+}$	-	Prussian blue	Prussian blue	4
TiO_2	-	TiO_2	Pt	5
CdS	-	RuO_2	Pt	6

have been made. Typical methods for the photoinduced cleavage of water systems are summarized in Table 3.1.

Kaneko et al. reported the photoinduced cleavage of water system of prussian blue as a catalyst for oxygen and hydrogen evolution using $Ru(bpy)_3^{2+}$ as a photosensitizer.[4] In this system, the ratio of the amount of oxygen to hydrogen evolution is 1 to 2, the stoichiometric ratio of water decomposition. The quantum yield is estimated to be about 10^{-2}, because the back electron transfer reaction rate is very rapid compared with hydrogen and oxygen evolution. Studies on photoinduced cleavage of water using the modified titanium oxide $(TiO_2)^{[5,7]}$ semiconductor[6] have been performed.

Green plants split water under irradiation by sunlight. The brief reaction mechanism of the green plant photosystem is shown in Scheme 3-4. Green plants extract electrons by water oxidation with the aid of light energy and the electrons are transported into the Calvin cycle through ferredoxin. The arrows in the scheme show the electron flow. By the addition of hydrogenase to extract electrons from the ferredoxin, the electrons are made to flow in the direction shown by the dotted arrow and hydrogen evolves through the reduction of protons. Photoinduced hydrogen evolution is reported through the irradiation of a system containing chloroplast-ferredoxin-hydrogenase. The main results are summarized in Table 3.2. The amount of hydrogen evolved per mg-chloroplast is *ca.* 10 mmol. Rao et al. reported 188 μmol of hydrogen evolution with 2 h irradiation.[8] In each system, however, the hydrogen evolution rate decreases with irradiation time, and it is difficult to obtain hydrogen continuously

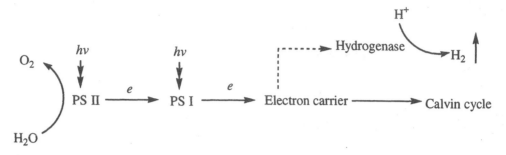

Scheme 3-4 Diagram of electron flows in photosystem.

Table 3.2 Photoinduced hydrogen evolution with the chloroplast-ferredoxin-
hydrogenase system

Hydrogen evolved	Chloroplast	Irradiation time	Ref.
0.25 μmol	0.8 mg	15 min	9
10 μmol	1 mg	1 h	10
60-70 μmol	1 mg	7-8 h	11
30-50 μmol	1 mg	1 h	8
(Maximum188 μmol/mg chloroplast/2 h)			

for a long time. The cause of the decrease is the simultaneous evolution of oxygen, which is a strong inhibitor of hydrogenase. To obtain hydrogen over a long period it is necessary to remove oxygen. One method is to consume oxygen by the addition of glucose and glucose oxidase. Another method is to develop a hydrogenase which is unaffected by oxygen. Kamen et al. found that hydrogenase from *Clostridium pasteurianum* when fixed on glass beads was more stable towards oxygen.[11] To avoid enzyme deactivation by oxygen, Greenbaum et al. used a flow system of compressed carrier gas (either helium or nitrogen) which continually purges the reaction cuvettes and electrolysis cell.[12] They measured the simultaneous photoproduction of hydrogen and oxygen in marine green algae. *Chlamydomonas* species (clone f-9), for example, has a steady state rate of hydrogen and oxygen production during irradiation with a stoichiometric ratio close to 2:1.

However, an effective photoinduced cleavage of water system using Scheme 3-2 has not yet been established. The photoinduced cleavage of water system is divided into the photoinduced hydrogen evolution system as shown in Scheme 3-5 (1) and the oxygen evolution system as shown in Scheme 3-5 (2), and each system has been investigated to establish the effective photoinduced cleavage of water.

Photoinduced cleavage of water will be established principally by the combination of hydrogen and oxygen evolution systems. Although such an effective system has not yet been established, many studies on photoinduced hydrogen evolution systems have been performed because hydrogen is an important future energy resource, as noted in section 3.1.

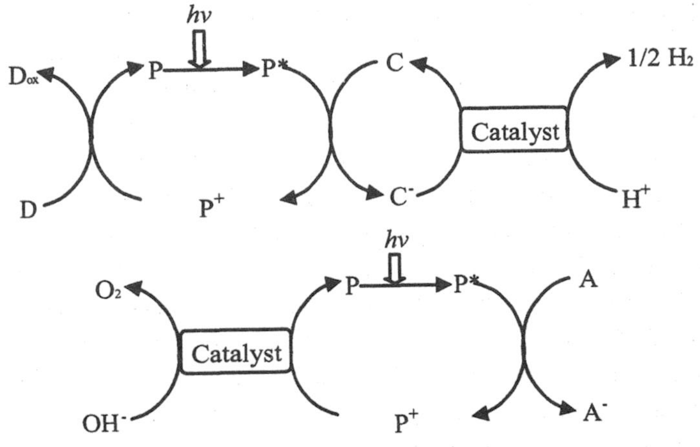

Scheme 3-5 Photoinduced hydrogen evolution (1) and oxygen evolution (2) system.
D: Electrondonor, P: Photosensitizer, C, A: Electron carrier.

3.3 Photoinduced Hydrogen Evolution in Homogeneous Four-component Systems

Photoinduced hydrogen evolution systems consisting of an electron donor (D), a photosensitizer (P), an electron carrier (C) and a catalyst as shown in Scheme 3-6 have been widely studied. Porphyrins are suitable photosensitizers for photoinduced hydrogen evolution. Hydrogenase and colloidal platinum are widely used as hydrogen evolution catalysts. As the reaction conditions are limited for enzyme catalysis, replacement of the enzyme hydrogenase by artificial heterogeneous or homogeneous catalyst has been undertaken. Hall et al. compared the efficiencies of platinum and hydrogenase and found that

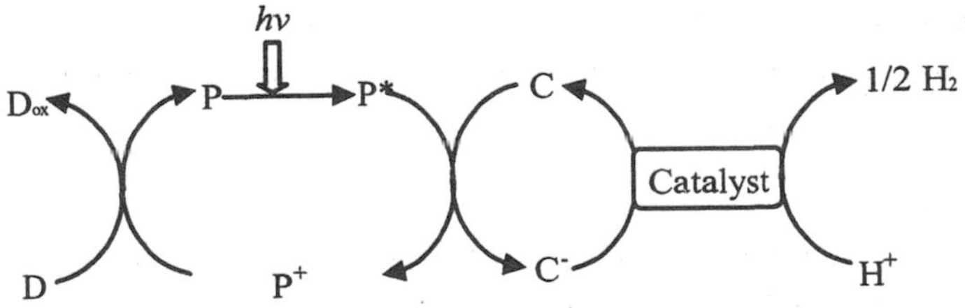

Scheme 3-6 Photoinduced hydrogen evolution system. D: Electron donor,
P: Photosensitizer, C: Electron carrier.

both lead to evolution of hydrogen at similar rates (7-9 μmol-H$_2$ h^{-1} mg-chlorophyll^{-1}) on illumination of isolated chloroplasts with water as the electron source and methyl viologen as an electron carrier.[7] The activities, the turnover numbers per active site, for colloidal platinum and hydrogenase were determined to be 3.16×10^{-24} and 2.83×10^{-21} mol min^{-1}, respectively. From these results it is apparent that hydrogenase is hundreds of times more active than colloidal platinum. The time dependences for photoinduced hydrogen evolution by hydrogenase and by colloidal platinum are shown in Fig. 3.1. In these experiments the concentration of the reduced form of methyl viologen is kept constant during the reaction to make sure that the rate-determining step of this reaction is the proton reduction by reduced methyl viologen. In the case of hydrogenase, hydrogen is evolved linearly with time and no deactivation of hydrogenase is observed within 24 h. On the other hand, the hydrogen evolution rate gradually decreases in the case of colloidal platinum. The rate decrease is considered to be caused by colloidal platinum coagulation and/or the decrease of the effective concentration of methyl viologen by hydrogenation with the hydrogen produced by colloidal platinum. Johansen et al. and Keller et al. reported that colloidal platinum was active for methyl viologen hydrogenation.[13,14] Typical photoinduced hydrogen evolution systems with porphyrins and hydrogenase are described below.

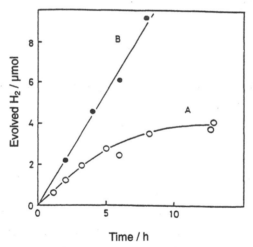

Fig. 3.1 Time dependence of photoinduced hydrogen evolution with the ZnTPPS-methyl viologen-RSH system catalyzed by hydrogenase (B) and colloidal platinum (A) at pH 7.0. (Reproduced with permission by K. Kalyanasundaram, B. Borgarello, G. Gratzel, *Helv. Chim. Acta*, **64** (1981) 362)

In 1975, Krasnovski et al. reported photoinduced hydrogen evolution with hydrogenase by the combination photosystem shown in Scheme 3-7.[15] In this reaction, NADH is used as the electron donor, chlorophyll (Chl) as photosensitizer, methyl viologen as electron carrier and hydrogenase as catalyst. As the chlorophyll is decomposed by the irradiation in the system, steady hydrogen evolution is not achieved.

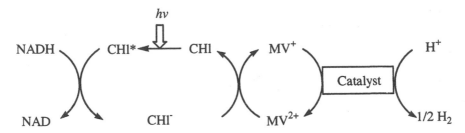

Scheme 3-7 Photoinduced hydrogen evolution with NADH-chlorophyll-methylviologen-hydrogenase system. (Reproduced with permission by A.A. Krasnovski, G.P. Brin, I.P. Gotov, V.P. Pschepkov, *Dokl. Acad. Nauk. SSSR.*, **225** (1975) 711)

Krasnovski et al. also reported photoinduced hydrogen evolution with hematoporphyrin (Scheme 3-8), the first report on photoinduced hydrogen evolution using porphyrin as a photosensitizer.[16] Since this report, extensive studies on photoinduced hydrogen evolution using porphyrins have been carried out. Typical photoinduced hydrogen evolution systems are summarized in Table 3.3.[16-22]

In the photoinduced hydrogen evolution with a four-component system, the reaction mechanisms are classified into two types: oxidative quenching and reductive quenching (Scheme 3-9).

The quenching reaction, whether the reaction proceeds via oxidation or reduction, depends on the structure and redox potential of the sensitizer. In the case of zinc porphyrins, the reductive

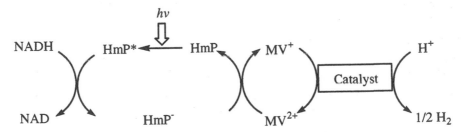

Scheme 3-8 Photoinduced hydrogen evolution with NADH-hematoporphyrin-methyl viologen-hydrogenase system. (Reproduced with permission by A.A. Krasnovski, G.P. Brin, U.V. Nikandrov, *Dokl. Acad. Nauk. SSSR.*, **228** (1975) 1214)

Table 3.3   Typical photoinduced hydrogen evolution systems.

Electron donor	Photosensitizer	Electron carrier	Catalyst for hydrogen evolution	Ref.
NADH	Hematoporphyrin	MV^{2+}	H_2ase	16
EDTA	Uroporphyrin	MV^{2+}	Pt colloid	17
RSH	ZnTPP	MV^{2+}	Pt colloid	18
RSH	ZnTPP	MV^{2+}	H_2ase	18
EDTA	ZnTMPyP	MV^{2+}	Pt colloid	19,20
RSH	ZnTPPS	MV^{2+}	Pt colloid, H_2ase	21,22

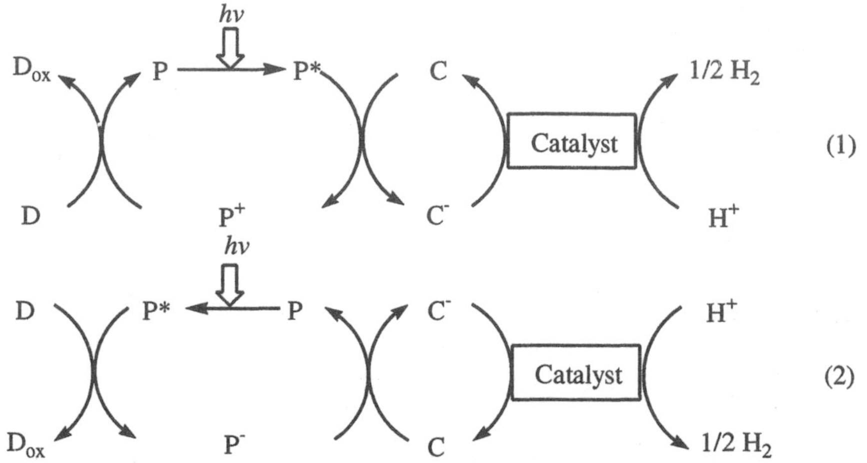

Scheme 3-9 Oxidative quenching (1) and reductive quenching (2)
in photoinduced hydrogen evolution.

quenching occurs when cationic zinc porphyrins, zinc tetrakis(4-methylpyridyl)porphyrin (ZnTMPyP) and their derivatives, are used. Under reductive quenching, zinc porphyrins become unstable by irradiation, for the intermediates are easily decomposed. On the other hand, oxidative quenching occurs when anionic porphyrins, zinc tetraphenylporphyrin tetrasulfonate (ZnTPPS), are used. A high yield of photoinduced hydrogen evolution is observed under oxidative quenching.

Some metalloporphyrins have recently been found to be efficient photosensitizers though ruthenium complexes have been employed exclusively as photosensitizers for the photoinduced hydrogen evolution.[22] Anionic metalloporphryins are especially superior as described above. ZnTPPS exhibited particularly high activity for the reaction as shown in Table 3.4, where the activity is compared by the initial photoreduction rate of methyl viologen. In this section, on the basis of laser flash photolysis and steady state irradiation, the reasons for the high activity of ZnTPPS for the photoreduction of methyl viologen are described. Furthermore, not only methyl viologen, the most popular electron carrier, but also zwitterionic 1,1'-dipropylviologen sulfonate (PVS), which has a different charge from methyl viologen, is used as an electron carrier and a comparison of the efficiency of the two viologens

Table 3.4 Activity of several photosensitizers for the photoreduction of methyl viologen

Photosensitizer	Activity (mol-MV$^+$ / mol-Photosensitizer)
Zn-TPPS$_3$$^{3-}$	93.2
Cd-TPPS$_3$$^{3-}$	43.8
Mn-TPPS$_3$$^{3-}$	24.9
Co-TPPS$_3$$^{3-}$	0
Cu-TPPS$_3$$^{3-}$	0
H$_2$-TPPS$_3$$^{3-}$	3.42
Zm-TPP	2.2
HmP	0.98
Ru(bpy)$_3$$^{2+}$	0.87

is made. Based on the results obtained, a reaction mechanism for the photoreduction of viologens is proposed.

3.3.1 Bipyridinium compounds as electron carriers

When an aqueous solution containing 2-mercaptoethanol as the electron donor and ZnTPPS and methyl viologen as electron carrier was irradiated, the growth of reduced methyl viologen with characteristic absorption bands at 395 and 605 nm is observed as shown in Fig. 3.2. The concentration of reduced methyl viologen increases with irradiation time and tends to reach a constant value. After 60 min irradiation the yield of reduced methyl viologen is *ca.* 70%. The reduction rates of methyl viologen (the slopes of the curves at the beginning of the reaction) depends on the concentrations of ZnTPPS and methyl viologen. The rate is proportional to the concentration of ZnTPPS as shown in Fig. 3.3. But an abnormal phenomenon is observed for the dependence of the initial rate on the concentration of methyl viologen as shown in Fig. 3.4. The reduction rate increases gradually with increase of methyl viologen concentration then decreases beyond the maximum value. When methyl viologen is added to the aqueous ZnTPPS solution, a change in the absorption spectrum of ZnTPPS is observed as shown in Figs. 3.5a and b, where the Soret band and Q band are shown, respectively. The absorption spectrum of ZnTPPS changed with isosbestic points when the amount of methyl viologen added increases, showing a complex formation between ZnTPPS and methyl viologen. Complex formations between porphyrins and methyl viologen have been reported in systems such as *meso*-tetra(4-sulfonatophenyl)porphyrin (TPPS)-methyl viologen,[23)] and ZnTPPS-methyl viologen.[24)] Similar change

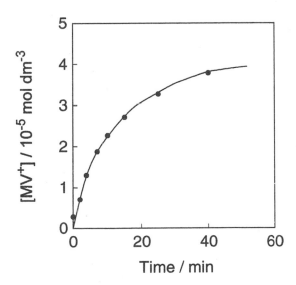

Fig. 3.2 Time dependence of reduced methyl viologen. Reaction condition: RSH (0.21 mol dm^{-3}), ZnTPPS (0.11 μmol dm^{-3}) and methyl viologen (6.1 x 10^{-5} mol dm^{-3}).

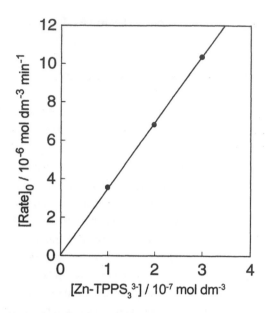

Fig. 3.3 Dependence of reduction rate on ZnTPPS concentration. Reaction condition: RSH (0.21 mol dm⁻³) and methyl viologen (1.2 x 10⁻⁴ mol dm⁻³).

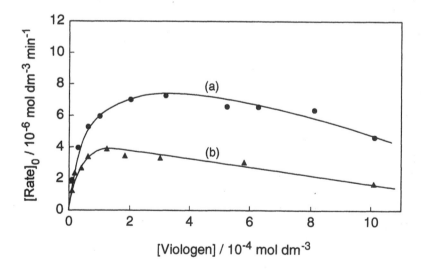

Fig. 3.4 Dependence of reduction rate on viologen concentration. Reaction condition: RSH (0.21 mol dm⁻³) and ZnTPPS (0.11 μmol dm⁻³). (a) PVS; (b) methyl viologen.

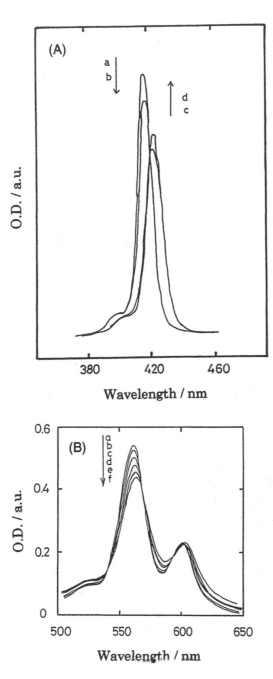

Fig. 3.5 Spectrum change of ZnTPPS (3.0 x 10⁻⁵ mol dm⁻³) by addition of methyl viologen. (A) Soret band; (B) Q band. Methyl viologen concentration: (a) 0, (b) 0.55 μmol dm⁻³, (c) 1.1 x 10⁻⁵ mol dm⁻³, (d) 4.9 x 10⁻⁵ mol dm⁻³, (e) 1.1 x 10⁻⁴ mol dm⁻³ and (f) 5.0 x 10⁻³ mol dm⁻³.

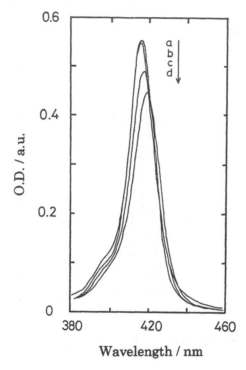

Fig. 3.6 Spectrum change of MgTPPS (1.3×10^{-6} mol dm^{-3}) by addition of methyl viologen. (a) 0, (b) 9.0×10^{-5} mol dm^{-3}, (c) 3.4×10^{-4} mol dm^{-3}, (d) 3.9×10^{-3} mol dm^{-3}.

in the absorption spectra is also observed when MgTPPS is used in place of ZnTPPS, as shown in Fig. 3.6, showing that the complex formation between porphyrin and methyl viologen does not depend on the central metal ions of the porphyrins.

When viologenpropylsulfonate (PVS) is used in place of methyl viologen, a similar change in the absorption spectrum change of ZnTPPS is observed, as shown in Fig. 3.7. However, a larger amount of PVS than methyl viologen is needed to observe the change in the absorption spectrum. This shows that methyl viologen forms the complex more easily than PVS. This is caused by the difference of the charge between methyl viologen and PVS, i.e., methyl viologen is positively charged while PVS is electrically neutral. No change in the absorption spectrum of ZnTPP is observed by adding 4,4'-bipyridine.

These results show that the complexes formed between ZnTPPS and viologens will be "ion pairs". The nature and the role of the complexes in photoreduction of viologens are described in a later section.

When an aqueous solution containing 2-mercaptoethanol, ZnTPPS, and PVS is irradiated, the growth of the reduced PVS is observed in the same way as methyl viologen, as shown in Fig. 3.8. The dependence of the reduction rate on the concentration of PVS is shown in Fig. 3.4. A similar phenomenon to the methyl viologen system is observed i.e., the reduction rate increases gradually with increase of PVS concentration and decreases beyond the maximum value. The degree of depression

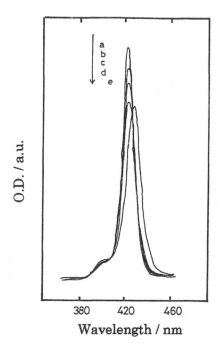

Fig. 3.7 Spectrum change of ZnTPPS (1.3 x 10^{-6} mol dm^{-3}) by addition of PVS. (a) 0, (b) 9.0 x 10^{-5} mol dm^{-3}, (c) 3.4 x 10^{-4} mol dm^{-3}, (d) 3.9 x 10^{-3} mol dm^{-3}.

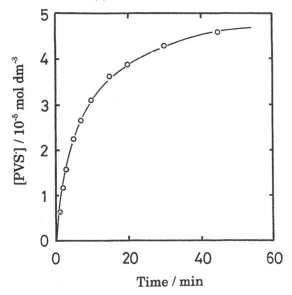

Fig. 3.8 Time dependence of reduced PVS. Reaction condition: RSH (0.21 mol dm^{-3}), ZnTPPS (0.11 μmol dm^{-3}) and PVS (6.1 x 10^{-5} mol dm^{-3}).

of the reduction rate is smaller for PVS than for methyl viologen. For example, the rate at 1.0×10^{-3} mol dm^{-3} is about one half the maximum value, while the rate at the same concentration of PVS is about two thirds the maximum value. The decrease of the reduction rate at the higher concentration of viologens described above is caused by the complex formation between ZnTPPS and viologens. The mechanism of the decrease of the reduction rate is described in detail in a later section.

Many types of bipyridinium salts can replace methyl viologen, and some of them are more suitable electron carriers than methyl viologen. Chemical structures of bipyridinium compounds are shown in Fig. 3.9. When the bipyridinium compounds are added to the ZnTPPS aqueous solution, the Soret band of ZnTPPS (418 nm) changes with the isosbestic point. This indicates that an electrostatic complex is formed between ZnTPPS and bipyridinium compounds. Table 3.5 shows the association constant (K_{aso}) of the complex formation between ZnTPPS and bipyridinium compounds. In the cases of methyl viologen and compound (B), K_{aso} is estimated to be 10^4 mol^{-1}dm^3. On the other hand, K_{aso} is estimated to be 1.0×10^3 mol^{-1} dm^3 in compounds (A), (C) and (D).[25] Thus these compounds do not form complexes easily compared with methyl viologen and (B). When an aqueous solution containing ZnTPPS, bipyridinium compounds, 2-mercaptoethanol and hydrogenase is irradiated, hydrogen evolution at a stationary rate is observed as shown in Fig. 3.10.[25] It is evident that all the compounds can serve as electron carriers for photoinduced hydrogen evolution. Hydrogen evolution rates of systems with compounds (A), (C) and (D) are much greater than those with methyl viologen. Redox potentials of methyl viologen, (A), (B), (C) and (D) are –0.44, –0.72, –0.37, –0.55 and –0.65 V (vs. NHE). As the redox potential of (B) is not sufficient to reduce water, the hydrogen evolution efficiency is lower than that of the other compounds.

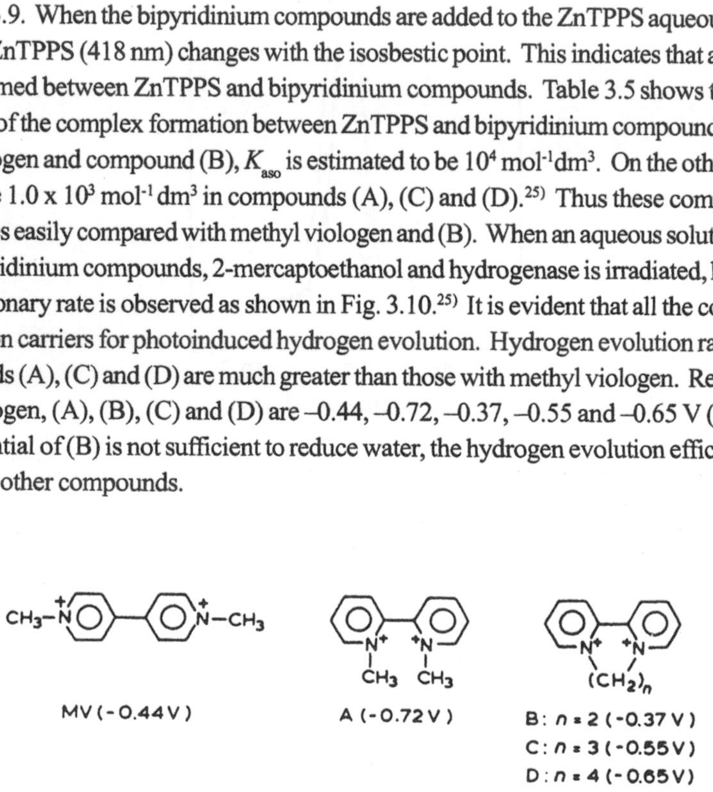

Fig. 3.9 Various bipyridinium salt compound as electron carrier (fingers in parentheses indicate the redox potentials vs. NHE).

Table 3.5 Association constant (K) of the complex formation between [Zn-TPPS$_3$]$^{3-}$ and bipyridinium compound

Bipyridinium compound	MV	A	B	C	D
K / 10^3 mol^{-1} dm^3	16 ± 2	1.7 ± 0.4	12 ± 1	3.4 ± 0.6	1.8 ± 0.2

(Reproduced with permission from *Inorg. Chem.* **24** (1985) 451, ACS)

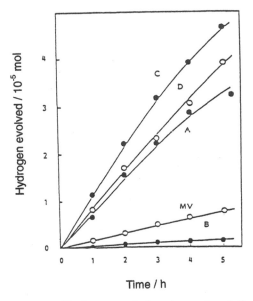

Fig. 3.10 Time dependence of hydrogen evolution. A sample solution containing RSH (0.2 mol dm^{-3}), ZnTPPS (0.7 mmol dm^{-3}), bipyridinium salt (1.3 x 10^{-4} mol dm^{-3}) and hydrogenase (0.5 ml) is irradiated by visible light at 30°C. Electron carriers labeled as in Fig. 3.9. (Reproduced with permission from *Inorg. Chem.* **24** (1985) 451, ACS)

A transient difference spectrum after the excitation of ZnTPPS in deaerated aqueous solution is shown in Fig. 3.11. The typical decay curves observed at 470 nm are shown in Fig. 3.12(A). The decay follows the first-order kinetics as shown in Fig. 3.12(C). This transient absorption is quenched efficiently by molecular oxygen as shown in Fig. 3.12(B); this is observed at 470 nm in aqueous solution saturated with air. Furthermore, the spectrum shown in Fig. 3.13 is similar to the triplet-triplet (T-T)

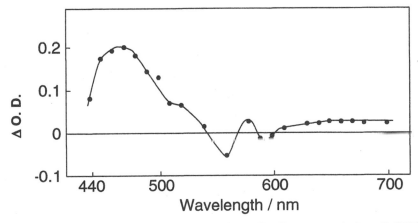

Fig. 3.11 Difference transient absorption spectra for ZnTPPS aqueous solution. ZnTPPS (3.0 x 10^{-5} mol dm^{-3}) in deaerated solution, observed at 30 μs after the excitation.

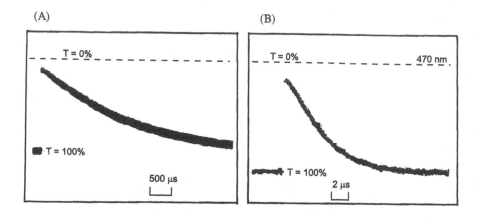

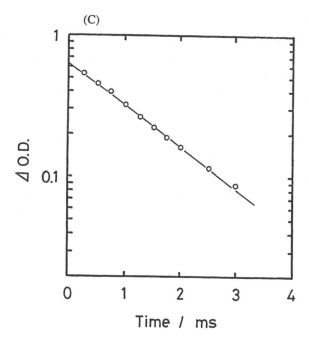

Fig. 3.12 Typical decays for T-T absorption of ZnTPPS. (A) dearated, (B) air saturated conditions, (C)
first order plot of (A).

absorption spectrum of porphyrins reported previously, which have a structure similar to that of
ZnTPPS.[26] Based on the results described above, it is concluded that the spectrum shown in Fig. 3.13
is the difference T-T absorption spectrum for the photoexcited triplet state of ZnTPPS.

The lifetime of the photoexcited triplet state of ZnTPPS is estimated to be 1.6 ms from Fig.

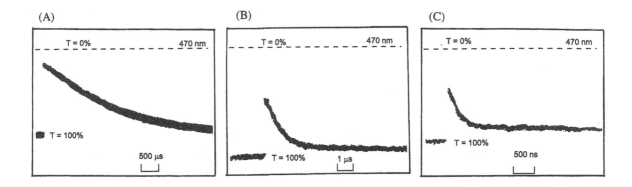

Fig. 3.13 Typical decays for T-T absorption of ZnTPPS. (A): ZnTPPS (3.0×10^{-5} mol dm^{-3}), (B): (A) + methyl viologen (7.6×10^{-5} mol dm^{-3}), (C): (A) + PVS (1.5×10^{-3} mol dm^{-3}).

3.12(A). Such a long lifetime is one of the reasons why ZnTPPS is a good photosensitizer for the photoreduction of viologens.

   Both methyl viologen and PVS quench the photoexcited triplet state of ZnTPPS efficiently, as shown in Fig. 3.13, in which the typical decays are shown. In both cases the decay curves consist of two parts, i.e., one component with a shorter lifetime and the other with a longer lifetime.

   The transient spectra of the sample solution containing ZnTPPS and PVS after the laser flash are shown in Fig. 3.14. At 2.5 μs after the excitation the absorptions of both species with shorter lifetime and longer lifetime are observed, but at 30 μs after the excitation the species with shorter lifetime disappeared and only the species with the longer lifetime remained. The transient spectrum (b) in Fig. 3.14 is attributed to the one electron oxidation product of ZnTPPS (ZnTPPS$^+$) and the one-electron reduced product of PVS (PVS$^-$), which are produced by the reaction as shown in following equation.

$$^3\text{ZnTPPS} + \text{PVS} \quad \rightarrow \quad \text{ZnTPPS}^+ + \text{PVS}^-$$

   The decay curves of the species with longer lifetime at 470 nm are shown in Figs. 3.15(A) and 3.16(A), corresponding to the systems containing methyl viologen and ZnTPPS, and PVS and ZnTPPS, respectively. In both cases, these decays follow second-order kinetics, as shown in Figs 3.15(B) and 3.16(B), attributed to the following reaction.

$$\text{ZnTPPS}^+ + \text{Viologen}^- \xrightarrow{k_q} \text{ZnTPPS} + \text{Viologen}$$

   The species which has the shorter lifetime is attributed to the photoexcited triplet state of

ZnTPPS, for the decays follow first-order kinetics in the cases of both methyl viologen and PVS, as shown in Figs. 3.16(A) and (B). To determine the rate constant for the quenching of the photoexcited triplet state of ZnTPPS (k_q in the following equation), the Stern-Volmer plots are shown in Figs. 3.17 and 3.18, corresponding to the methyl viologen - ZnTPPS, and the PVS - ZnTPPS systems, respectively.

$$^3ZnTPPS \ + \ Viologen \quad \rightarrow \quad [\ ZnTPPS \ \cdots \ Viologen\ ]$$

The values of k_q obtained from the Stern-Volmer plots are summarized in Table 3.6. The value of k_q for methyl viologen is about seven times greater than that for PVS, both of which are close to the diffusion-controlled rate constants under the reaction conditions. This is due to the difference in the electrostatic interaction between the photoexcited triplet state of ZnTPPS and viologens. Methyl viologen is positively charged and PVS is electrically neutral, while the photoexcited triplet state of ZnTPPS is negatively charged and PVS is electrically neutral, while the photoexcited triplet state of ZnTPPS is negatively charged. Thus, the electrostatic interaction between the photoexcited triplet state of ZnTPPS and methyl viologen is greater than that between the photoexcited triplet state of

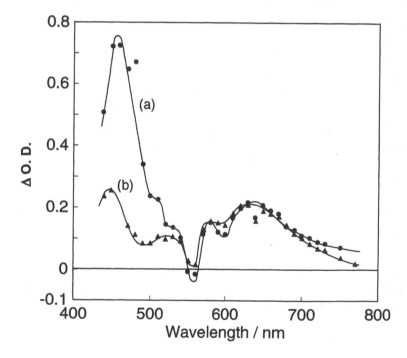

Fig. 3.14 Difference transient absorption spectra for ZnTPPS-PVS solution. ZnTPPS (3.0 x 10⁻⁵ mol dm⁻³) and PVS (1.5 x 10⁻⁴ mol dm⁻³) in deaerated solution observed at 2.5 μs (a) and 30 μs (b) after the excitation.

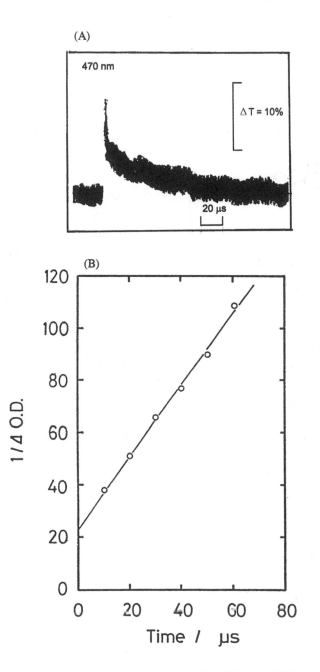

Fig. 3.15 Typical decays for longer lifetime species observed at 470 nm and (B) second order plot for the decay curve (A). ZnTPPS (3.0×10^{-5} mol dm^{-3}) and methyl viologen (7.6×10^{-5} mol dm^{-3}) in deaerated solution.

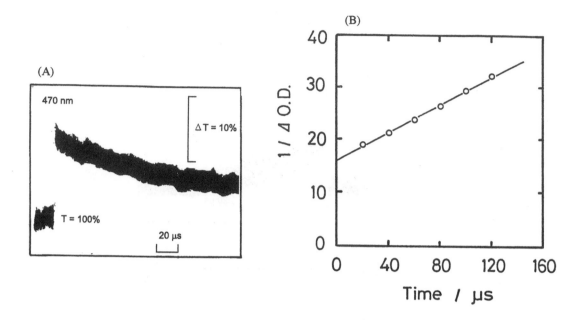

(A)

470 nm

ΔT = 10%

T = 100%

20 μs

(B)

Fig. 3.16 Typical decays for longer lifetime species observed at 470 nm and (B) second order plot for the decay curve (A). ZnTPPS (3.0×10^{-5} moldm^{-3}) and PVS (1.5×10^{-3} mol dm^{-3}) in deaerated solution.

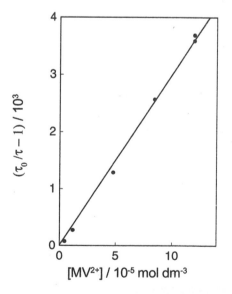

Fig. 3.17 Stern-Volmer plot for the ZnTPPS-methyl viologen system.

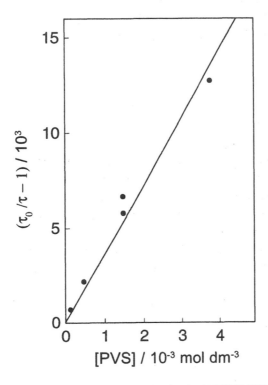

Fig. 3.18 Stern-Volmer plot for the ZnTPPS-PVS system.

Table 3.6 Quenching rate constants of ^{3}Zn-TPPS$_3^{3-}$ by several quenchers

Quencher	k_q/ mol^{-1} dm^3 s^{-1}	$k_q^{\mu=0}$ / mol^{-1} dm^3 s^{-1}
MV^{2+}	$(1.5\pm0.3) \times 10^{10}$	1.5×10^{10}
PVS	$(2.3\pm0.3) \times 10^{9}$	2.6×10^{9}
RSH	$(4.7\pm1.9) \times 10^{4}$	-

ZnTPPS and PVS. As the redox potentials of methyl viologen and PVS are –0.44 [27] and –0.41 V (vs NHE),[28] respectively, the thermodynamic driving force for the quenching reaction is almost the same. Therefore the difference in the value of k_q is caused mainly by the electrostatic interaction.

The photoexcited triplet state of ZnTPPS is quenched also by 2-mercaptoethanol, but the quenching rate constant for 2-mercaptoethanol is about five orders of magnitude smaller than that for viologens, as shown in Table 3.6. Therefore, it is concluded that the quenching of the photoexcited triplet state of ZnTPPS by viologen is dominant, or the oxidative quenching of the photoexcited triplet state of ZnTPPS is more favorable.

As the quenching reactions mentioned above are ionic reactions, the value of k_q is expected to be strongly dependent on the ionic strength. If the Debye Limiting Rule is valid under the reaction conditions the following, Debye-Bronsted equation is obtained.

$$\log k = \log k_0 + 2AZ_AZ_B\mu^{1/2}$$

where k and k_0 are the rate constants at ionic strength of μ and zero, respectively, and Z_A and Z_B are the charges of reactants A and B, respectively.

The relations between $\log k_q$ and $\mu^{1/2}$ for the ZnTPPS-methyl viologen, and ZnTPPS-PVS systems are shown in Figs. 3.19 and 3.20, respectively. As expected from the above equation, for the ZnTPPS-methyl viologen system the linear relation with negative slope is obtained, while for the Zn-TPPS-PVS system $\log k_q$ is independent of $\mu^{1/2}$. This means that the above equation is valid for these systems. From the results shown in Figs. 3.21 and 3.22, the values of k_q at $\mu=0$ by extrapolation of the straight lines are calculated as listed in Table 3.6.

The photoexcited triplet state of ZnTPPS is quenched efficiently by the viologens described above, and at the same time, the formation of reduced viologens is observed. The formation of reduced viologens, MV^+ and PVS^- which have a characteristic absorption band at 605 nm after laser flash are shown in Figs. 3.21(B) and 3.22(B), respectively. The formation of reduced viologens corresponds to the photoexcited triplet state of ZnTPPS observed at 470 nm, shown in Figs. 3.21(A) and

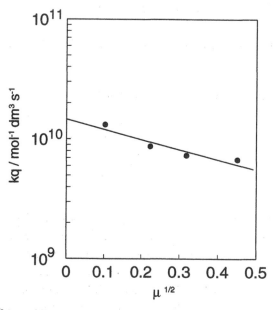

Fig. 3.19 Dependence of quenching rate constant on the ionic strength. ZnTPPS (3.0×10^{-5} mol dm^{-3}) and methyl viologen (4.9×10^{-5} mol dm^{-3}) in deaerated solution. Ionic strength is changed by NaCl solution.

3.22(A). This shows that the following reaction occurs.

$$^3\text{ZnTPPS} + \text{Viologen} \xrightarrow{k_q} \text{ZnTPPS}^+ + \text{Viologen}^-$$

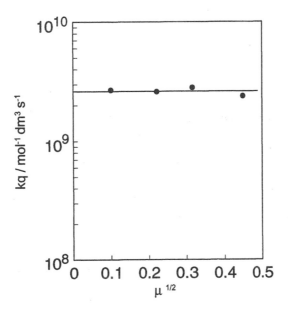

Fig. 3.20 Dependence of quenching rate constant on the ionic strength. ZnTPPS (3.0×10^{-5} mol dm^{-3}) and PVS (4.9×10^{-4} mol dm^{-3}) in deaerated solution. Ionic strength is changed by NaCl solution.

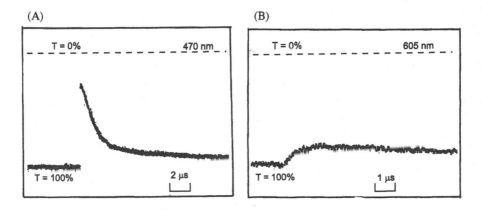

Fig. 3.21 Decay of T-T absorption of ZnTPPS (A) and formation of the absorption for the reduced methyl viologen (B). ZnTPPS (3.0×10^{-5} mol dm^{-3}) and methyl viologen (4.9×10^{-5} mol dm^{-3}) in deaerated solution.

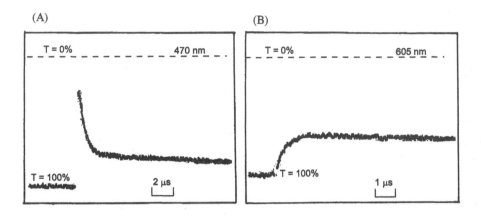

Fig. 3.22 Decay of T-T absorption of ZnTPPS (A) and formation of the absorption for the reduced PVS
(B). ZnTPPS (3.0×10^{-5} mol dm^{-3}) and PVS (4.9×10^{-4} mol dm^{-3}) in deaerated solution.

In Figs. 3.23(A) and 3.24(A), the decay curves observed at 605 nm are shown. The decay
curves follow second-order kinetics, as shown in Figs. 3.23(B) and 3.24(B). These results are explained by the following equation.

$$\text{ZnTPPS}^+ + \text{Viologen}^- \rightarrow \text{ZnTPPS} + \text{Viologen}$$

The transient absorption at 605 nm contains not only the absorption of reduced viologen but
also that of ZnTPPS$^+$. As the absorption of ZnTPPS$^+$ at 605 nm is small,[29] the transient absorption of
ZnTPPS$^+$ at 605 nm is negligible.

The relative rate constants, k_b, obtained from the results described above are listed in Table
3.7. The value for methyl viologen is about seven times greater than that for PVS.

Under the reaction conditions, the efficiency of the quenching reaction, which is defined by the
following equation, is 100% for both methyl viologen and PVS. This shows that all of the photoexcited
porphyrins react with viologens and that the difference in k_q values does not affect the reaction rate
under steady state irradiation.

$$\eta_q = k_q [\text{V}^{2+}] / (k_0 + k_q [\text{V}^{2+}])$$

where [V^{2+}] denotes the concentration of the bipyridinium salt and k_0 is the rate constant for
the spontaneous deactivation of the ZnTPPS triplet state.

The difference in reduction rates under steady state irradiation is explained by k_b values, i.e.,
k_b in the ZnTPPS-PVS system is smaller than that in the ZnTPPS-methyl viologen system, so the

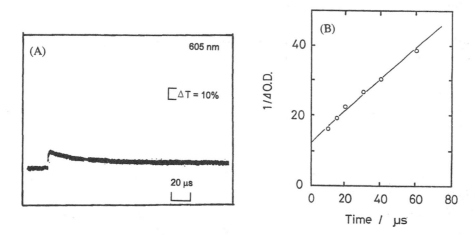

Fig. 3.23 Decay of the absorption for the reduced methyl viologen (A) and second order plot of (A).
ZnTPPS (3.0×10^{-5} mol dm^{-3}) and methyl viologen (4.9×10^{-5} mol dm^{-3}) in deaerated solution.

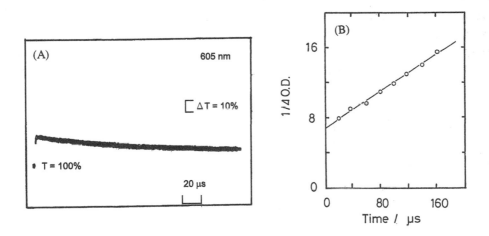

Fig. 3.24 Decay of the absorption for the reduced PVS(A) and second order plot of (A). ZnTPPS (3.0×10^{-5} mol dm^{-3}) and PVS (1.1×10^{-5} mol dm^{-3}) in deaerated solution.

Table 3.7 Relative rate constants for back electron transfer reactions

Reaction	k_b / (relative value)
Zn- TPPS$_3{}^{3-}$ + MV$^+$	100
Cd- TPPS$_3{}^{3-}$ + PVS$^-$	14

reduction rate of PVS is higher than that of methyl viologen under steady state irradiation.

The quenching of the photoexcited triplet state of ZnTPPS and the formation of ZnTPPS⁺ and reduced viologens proceed according to the following scheme, which is considered to be the general reaction scheme of photoinduced electron transfer.[30]

$$P^* + Q \; \overset{\rightarrow}{\leftarrow} \; [P \cdots Q]^* \; \rightarrow \; [P^+ \cdots Q^-] \; \begin{smallmatrix} \overset{k_1}{\nearrow} \; P^+ + Q^- \\ \underset{k_2}{\searrow} \; P + Q \end{smallmatrix}$$

The photoexcited triplet state of ZnTPPS forms the encounter complex with a quencher (viologens), and electron transfer takes place within the solvent cage. Then the separated ion pairs are formed (k_1 in the above scheme) or the recombination of ions within the solvent cage takes place (k_2 in the above scheme). The efficiency of the charge separation, i.e., the quantum yield for the formation of reduced viologen against the photoexcited triplet state of ZnTPPS formed (Φ), influences the efficiency of the photoreduction of viologens under steady state irradiation.

The relative quantum yield for the formation of reduced viologens is obtained from the plots shown in Fig. 3.25. In Fig. 3.25, $\Delta O.D.^{470}$ is the difference absorbance observed at 470 nm immediately after the excitation, which corresponds to the concentration of the photoexcited triplet state of ZnTPPS formed, and $\Delta O.D.^{605}$ is the maximum difference absorbance observed at 605 nm, and ε^{605} is the absorption coefficient of reduced methyl viologen ($\varepsilon^{605} = 1.1 \times 10^4 \, mol^{-1} dm^3 cm^{-1}$) or of reduced PVS ($\varepsilon^{605} = 1.3 \times 10^4 \, mol^{-1} dm^3 cm^{-1}$), and $\Delta O.D.^{605} / \varepsilon^{605}$ corresponds to the concentration of reduced

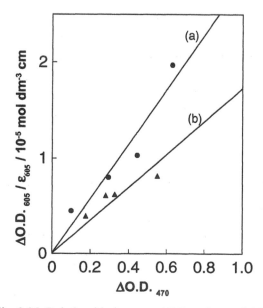

Fig. 3.25 Relationship between $\Delta O.D._{605} / \varepsilon_{605}$ and $\Delta O.D._{470}$.

viologen formed. Compared with the slopes of the straight lines, the relative value of Φ for the ZnTPPS-PVS system is obtained to be about 1.6 (when $\Phi = 1.0$ for the ZnTPPS-methyl viologen system).

This result is also explained by the electrostatic interaction between ZnTPPS and viologens. Electrostatic repulsion exists between ZnTPPS$^+$ and PVS$^-$ since both ZnTPPS$^+$ and PVS$^-$ are negatively charged. The reduced methyl viologen is positively charged, so the recombination of ions within the solvent cage (k_1 in the above scheme) takes place more easily. Therefore, the value of Φ for PVS is larger than that for methyl viologen.

The previous section describes the complex formation that takes place between ZnTPPS and viologens. In this section more quantitative analysis is carried out using a laser flash photolysis.

Figure 3.26(A) shows a typical decay monitored at 470 nm (the peak wavelength of the T-T absorption band due to the photoexcited triplet state of ZnTPPS) after a laser flash. In the presence of methyl viologen the decay rate of the transient spectrum measured at 470 nm increased, as shown in Figs. 3.26(B), (C) and (D). The absorption due to the reduced methyl viologen measured at 605 nm increased initially then decreased as shown in Figs. 3.26(E), (F) and (G). The absorbance of the triplet spectrum at 470 nm measured immediately after the laser flash decreased with the increase in methyl viologen concentrations (Figs. 3.26(A),(B) and (C)), and the T-T absorption is not detected at concentrations higher than 5.0×10^{-3} mol dm^{-3}. As a consequence of the decrease in the absorbance at 470 nm, the concentration of reduced methyl viologen formed after the laser flash decreased, and formation of reduced methyl viologen is not observed in the absence of the T-T absorption, as shown in Fig. 3.26(G). These results are expressed in the following reaction scheme, suggesting a 1 : 1 complex (PQ) formation between ZnTPPS (P) and methyl viologen (Q),

$$
\begin{aligned}
\mathrm{P} \;+\; \mathrm{Q} &\;\rightleftarrows\; \mathrm{PQ} \\
\mathrm{P} &\;\rightarrow\; {}^3\mathrm{P}* \\
{}^3\mathrm{P}* \;+\; \mathrm{Q} &\;\rightarrow\; \mathrm{P}^+ \;+\; \mathrm{Q}^- \\
\mathrm{P}^+ \;+\; \mathrm{Q}^- &\;\rightarrow\; \mathrm{P} \;+\; \mathrm{Q} \\
\mathrm{PQ} \;\rightarrow\; \mathrm{PQ}* &\;\rightarrow\; \mathrm{PQ}
\end{aligned}
$$

where K is the association constant.

According to such a reaction scheme the concentration of PQ increases with increase in Q concentration. The absorbance of the T-T absorption at 470 nm is considered to be proportional to the concentration of uncomplexed ZnTPPS on the assumption that the photoexcited complex (PQ*) is converted into PQ very rapidly. Therefore, the following reaction is obtained.

$$ A_0 / A = 1 + K[Q] $$

where A_0 and A are the absorbance at 470 nm measured immediately after the laser flash in the absence and in the presence of methyl viologen, respectively.

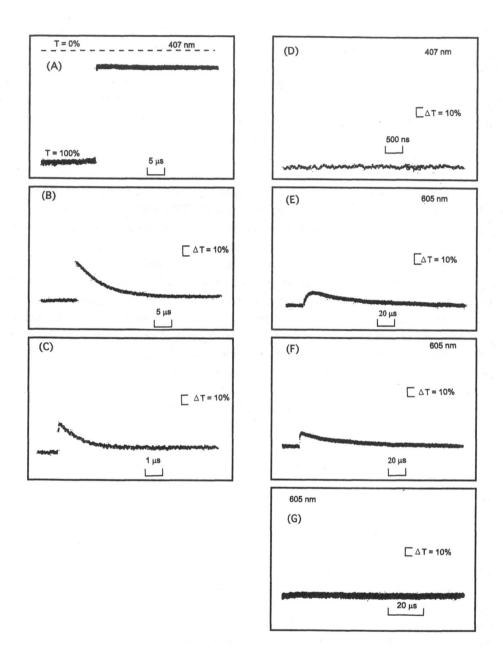

Fig. 3.26 Typical decays for T-T absorption of ZnTPPS and for absorption of the reduced methyl viologen. ZnTPPS (3.0×10^{-5} mol dm^{-3}), methyl viologen (a) 0, (b) 1.2×10^{-5} mol dm^{-3}, (c) 4.9×10^{-5} mol dm^{-3}, (d) 5.0×10^{-3} mol dm^{-3}, (e) 1.2×10^{-5} mol dm^{-3}, (f) 4.9×10^{-5} mol dm^{-3}, (g) 5.0×10^{-3} mol dm^{-3}; (a)-(d) λ_{obs}=470 nm; (e)-(g) λ_{obs}=605 nm, in deaerated solution.

As shown in Fig. 3.27, the fairly linear relationship between A_0 / A and [Q] indicates that the above proposed scheme is adequate. From the slope of the straight line, K is obtained as 1.7×10^4 mol^{-1} dm^3. When PVS is used in place of methyl viologen, a similar linear relationship is also obtained, as shown in Fig. 3.28, and K is estimated to be 7.3×10^2 mol^{-1} dm^3. The difference in the K values between PVS and methyl viologen is also explained by the difference in electrostatic force between ZnTPPS and viologen.

The difference between the redox potential of methyl viologen (-0.44 V) and that of PVS (-0.41 V) is so small that the large K value of methyl viologen compared with PVS is mainly due to the electrostatic effect. Since the positively charged methyl viologen is more favorable to associate nega-

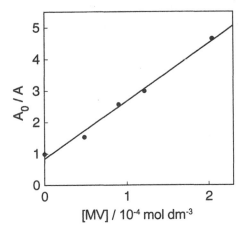

Fig. 3.27 Relationship between A_0 / A and the concentration of methyl viologen.

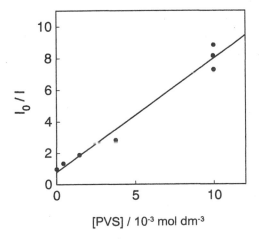

Fig. 3.28 Relationship between I_0 / I and the concentration of PVS.

tively charged ZnTPPS, the large K detected for methyl viologen compared with PVS is plausible.

Since the complexes formed between ZnTPPS and viologens (PQ) do not produce separated ion pairs, PQ is not favorable for the photoreduction of viologens. Therefore it is concluded that only uncomplexed ZnTPPS acts as a photosensitizer for the photoreduction of viologens.

The parameters obtained, k_q, k_b, Φ and K are summarized in Table 3.8. The fact that the reduction rate of PVS under steady state irradiation is greater than that of methyl viologen is explained by the difference in k_b, Φ and K values, i.e., the Φ value of PVS is greater than that of methyl viologen, and the k_b and K values of PVS are smaller than that of methyl viologen. When the K value is large, PQ, which is inactive for the photoreduction of viologens, is in high concentration. The concentration of uncomplexed ZnTPPS is greater in the ZnTPPS-PVS system than that in the ZnTPPS-methyl viologen system at the same concentration of viologens. Therefore the reduction rate of PVS is greater than that of methyl viologen under steady state irradiation. The following reaction mechanism is proposed for the photoreduction of viologens with ZnTPPS in the presence of 2-mercaptoethanol.

$$
\begin{aligned}
P &\rightarrow {}^3P* \\
{}^3P* &\rightarrow P \\
P + Q &\rightarrow PQ \\
{}^3P* + Q &\rightarrow P^+ + Q^- \\
[P^+ \cdots Q^-] &\rightarrow P + Q \\
[P^+ \cdots Q^-] &\rightarrow P^+ + Q^- \\
P^+ + Q &\rightarrow P + Q \\
PQ &\rightarrow PQ* \rightarrow PQ \\
P^+ + RSH &\rightarrow P + (RSH)_{ox}
\end{aligned}
$$

where P, P$^+$, Q, Q$^-$ are ZnTPPS, ZnTPPS$^+$, PVS or methyl viologen, and PVS$^-$ or reduced methyl viologen, respectively, and k_0, k_1, k_2 and k_b are rate constants, K is the association constant, I_0 is the intensity of the light, and Φ_0 is the quantum yield for the formation of the photoexcited triplet state.

On the basis of the above reaction mechanism, the following rare expression is derived by the use of the steady state approximation for [^{3}P*], [P$^+$] and [P$^+$Q$^-$].

$$
\begin{aligned}
V = d[Q^-]/dt &= I_0\Phi_0\Phi[P](k_q[Q]/(k_0+k_q[Q]))(k_3[RSH]/(k_b[Q^-]+k_3[RSH])) \\
&= I_0\Phi_0\Phi([P]_0/(1+K[Q]))(k_q[Q]/(k_0+k_q[Q]))(k_3[RSH]/(k_b[Q^-]+k_3[RSH]))
\end{aligned}
$$

Table 3.8 Obtained parameters by the laser flash photolysis

Quencher	k_q / mol^{-1} dm^3 s^{-1}	k_b / (relative value)	$\Phi_{rel.}$	K / mol^{-1} dm^3
MV^{2+}	$(1.5\pm0.3)\times10^{10}$	100	1.0	$(1.7\pm0.2)\times10^4$
PVS	$(2.3\pm0.3)\times10^9$	14	1.6	$(7.3\pm0.6)\times10^2$

where Φ means $k_2 / (k_1 + k_2)$ and $[P]_0$ is the concentration of ZnTPPS added.

Since $[Q^-]$ is neglected at the initial stage of the reaction, the following equation is obtained.

$$V_0 = I_0 \Phi_0 \Phi([P]_0 / (1 + K[Q]))(k_q[Q] / (k_0 + k_q[Q]))$$

where V_0 is the initial rate.

$$1 / V_0(1 + K[Q]) = (1 + k_0 / k_q[Q]) / I_0 \Phi_0 \Phi[P]_0$$

According to above equation, $1 / V_0(1 + K[Q])$ should be linear in relation to $1 / [Q]$. A good linear relation for both systems is obtained as shown in Figs. 3.29 and 3.30.

Under the reaction conditions of photoinduced hydrogen evolution, electron carriers with low redox potentials will be favorably photoinduced. The rate order, however, does not always coincide with the order of the redox potentials of the bipyridinium salts when ZnTPPS is used as a photosensitizer. The rate difference may depend on the efficiency of ion separation after electron transfer from the photoexcited ZnTPPS to the bipyridinium salts. The rates depend strongly on the molecular structure of the bipyridinium salts. The quenching rate constants are determined from the dependence of lifetime upon various bipyridinium salt concentrations (Table 3.9). These values are close to that for the diffusion-controlled process.

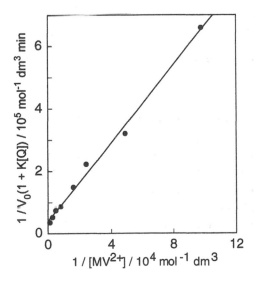

Fig. 3.29 Relationship between $1/ V_0 (1 +K[Q])$ and $1/ [MV^{2+}]$.

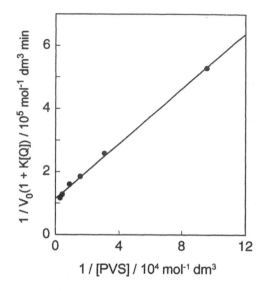

Fig. 3.30 Relationship between $1/V_0 (1 + K[Q])$ and $1/[PVS]$.

Table 3.9 Kinetic parameters obtained from laser flash photolysis of $ZnTPPS_3$-bipyridinium salt system

	MV	A	B	C	D
k_q (10^9 mol^{-1}dm^3 s^{-1})	15 ± 3	25 ± 18	99 ± 9	87 ± 13	53 ± 12
k_b (mol^{-1}dm^3 s^{-1})	100	21	38	18	17

(Reproduced with permission from *Inorg. Chem.*, **24** (1985) 451, ACS)

The relative back-reaction rate constants (k_b) are determined from the second-order plot (linear plot, ΔOD^{-1} versus time). k_b values depend strongly on the type of bipyridinium salt. When bipyridinium salts having smaller k_b values are used, higher hydrogen evolution rates are observed. Because compound B does not have a redox potential high enough to reduce protons and evolve hydrogen, the hydrogen evolution rate is very low, although the k_b value is also relatively small. From these results it is apparent that the back reaction (recombination reaction) plays a role in the photoinduced hydrogen evolution reaction in these systems.

3.3.2 Cytochrome c_3 as an electron carrier

Cytochrome c_3 is a natural electron carrier and a suitable substrate for hydrogenase, and is expected to be an efficient electron carrier for photoinduced hydrogen evolution. Photoreduction of cytochrome c_3 by irradiation of the system containing photosensitizer, cytochrome c_3 and triethanolamine as an electron donor, and hydrogen evolved catalyzed by hydrogenase are described.[31] When an aqueous solution containing ZnTPPS, cytochrome c_3 and triethanolamine is irradiated, the spectrum of cytochrome c_3 changes as shown in Fig. 3.31. The spectrum with 419 nm absorption attributed to an oxidized form of cytochrome c_3 decreases and a spectrum of 432 nm absorption appears. As reduction of cytochrome c_3 with dithionite yielded the same spectrum, the chemical species with 432 nm absorption can be attributed to a reduced form of cytochrome c_3. The concentration of this species increases rapidly at the beginning of the reaction and reaches a constant value. After 2 h irradiation almost all of the cytochrome c_3 exists in the reduced form. Aeration of the product leads to immediate reconversion to the starting compound, the oxidized form of cytochrome c_3, without loss. When hydrogenase is added to the system containing ZnTPPS, cytochrome c_3 and triethanolamine and the mixture is irradiated with sunlight for 6 h, hydrogen evolution is observed.

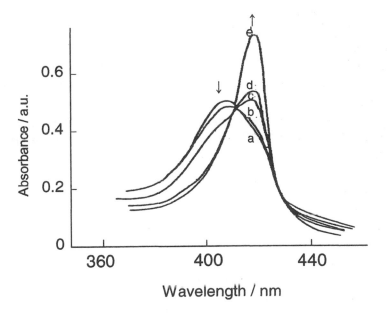

Fig. 3.31 Change of spectrum with irradiation time. The solution (pH 7.0) containing ZnTPPS, cytochrome c_3 and triethanolamine is irradiated for (a) 0, (b)6, (c)11, (d) 52 and (e) 117 min. (Reproduced with permission from *Chem. Lett.,* (**1982**) 187)

3.3.3 Reduction of NADP

In the systems shown in Scheme 3-6, new reactions (the reduction of substrates in place of proton) proceed by replacing the catalyst. The new catalyst should accept electrons from the reduced methyl viologen. Ferredoxin NADP reductase (FNR) catalyzes the reduction of NADP with the reduced methyl viologen, which acts as an artificial electron donor.[32] Thus, the photochemical reduction systems of NADP shown in Scheme 3-10 are achieved using FNR as a catalyst. In this section, photoreduction of NADPH using the system containing ZnTPPS, methyl viologen and FNR is described.

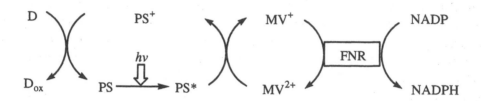

Scheme 3-10 Photoreduction of NADP with electron donor (D)-photosensitizer (PS)-methyl
viologen system.

As described above, ZnTPPS is a good photosensitizer for the photoreduction of methyl viologen. ZnTPPS is applied to the photoreduction system of NADP as photosensitizer. When a sample solution containing 2-mercaptoethanol, ZnTPPS, methyl viologen, FNR and NADPH is irradiated by visible light, the formation of NADPH (monitored at 340 nm, which corresponds to NADPH) is observed, as shown in Fig. 3.32. The dependence of the amount of NADPH on irradiation time is shown in Fig. 3.33. The amount of NADPH increases linearly with irradiation time while the conversion of NADP is low, then the reaction rate decreases gradually. In the case of the system shown in Fig. 3.33(d), the conversion of NADP is 64% after 2 h irradiation.

As shown in Fig. 3.33(a)-(d), the initial reduction rate of NADPH increases with the increase in the concentration of ZnTPPS. The dependence of the initial rate on the concentration of ZnTPPS is shown in Fig. 3.34. The initial rate increases linearly with increase in the concentration of ZnTPPS. Deviation, however, from the initial slope is observed at higher concentrations of ZnTPPS. This is caused by the saturation of the absorption of ZnTPPS for the Soret band.

The initial reduction rate of NADP is independent of the concentration of FNR, even at higher concentrations of ZnTPPS. This shows that the rate-limiting step of the reaction is the photoreduction of methyl viologen.

When compound C (the structure is shown in Fig. 3.11) is used in place of methyl viologen, reduction of NADP to NADPH is also observed. The dependence of the initial reduction rate on the concentration of ZnTPPS is shown in Fig. 3.35. The initial reduction rate in the case of C is greater than that in the case of methyl viologen.

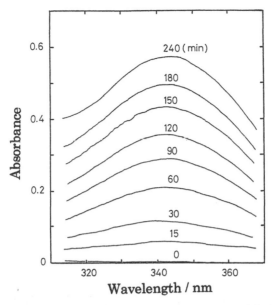

Fig. 3.32 Spectra change during steady state irradiation. RSH (0.24 mol dm^{-3}), ZnTPPS (1.9 x 10^{-7} mol dm^{-3}), methyl viologen (1.7 x 10^{-4} mol dm^{-3}), NADP (6.7 x 10^{-4} mol dm^{-3}) and FNR (0.18 unit ml^{-1}), pH 7.2.

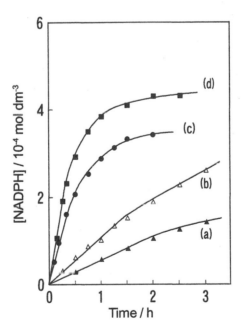

Fig. 3.33 Time dependence of the amount of NADPH. RSH (0.24 mol dm^{-3}), methyl viologen (1.7 x 10^{-4} mol dm^{-3}), NADP (6.7 x 10^{-4} mol dm^{-3}) and FNR (0.18 unit ml^{-1}), pH 7.2. ZnTPPS (a) 9.5 x 10^{-8}, (b) 1.9 x 10^{-7}, (c) 9.5 x 10^{-7} and (d) 1.9 x 10^{-6} mol dm^{-3}.

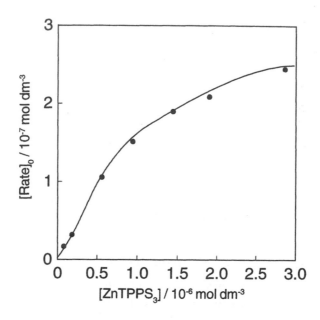

Fig. 3.34 Dependence of the initial rate on the concentration of ZnTPPS. RSH (0.24 mol dm⁻³),
methyl viologen (1.7 x 10⁻⁴ mol dm⁻³), NADP (6.7 x 10⁻⁴ mol dm⁻³) and FNR (0.18 unit
ml⁻¹), pH 7.2.

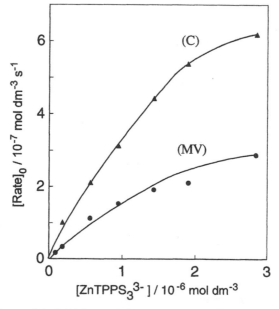

Fig. 3.35 Dependence of the initial rate on the concentration of ZnTPPS. RSH (0.24 mol dm⁻³), MV
or C (1.7 x 10⁻⁴ mol dm⁻³), NADP (6.7 x 10⁻⁴ mol dm⁻³) and FNR (0.18 unit ml⁻¹), pH 7.2.

At low concentrations of ZnTPPS, the absorption of NADPH increases gradually and then stops. At higher concentrations of ZnTPPS, even though the initial reduction rate is high, the absorption of NADPH disappears and a new absorption band appears below 320 nm by prolonged irradiation. When a higher concentration of ZnTPPS is used, more rapid disappearance of the absorption of NADPH is observed. When disappearance of the absorption of NADPH is observed, bleaching of the Soret band of ZnTPPS is also observed. This bleaching of the absorption of ZnTPPS is caused by the reductive decomposition of ZnTPPS. A similar bleaching of porphyrin has been reported previously, i.e., when ZnTMPyP is irradiated in the presence of EDTA, irreversible reduction of the porphyrin occurs and bleaching of the absorption of the porphyrin is observed.[26]

Since NADPH-requiring reductases catalyze the asymmetric reduction of substrates in the presence of NADPH, asymmetric reduction is accomplished in the system shown in Scheme 3-11 by adding the reductases to obtain valuable chiral products. The photochemical hydrogenation of ketone is accomplished by adding alcohol dehydrogenase to the photoreduction system of NADP.

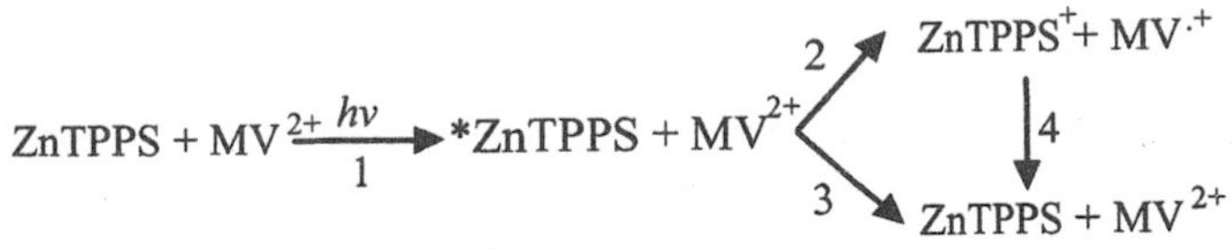

$$ZnTPPS + MV^{2+} \xrightarrow[1]{hv} *ZnTPPS + MV^{2+} \begin{array}{c} \nearrow^{2} ZnTPPS^{+} + MV^{\cdot+} \\ \downarrow 4 \\ \searrow_{3} ZnTPPS + MV^{2+} \end{array}$$

Scheme 3-11 Mechanism of methyl viologen photoreduction.

3.3.4 Effect of micelles

Many types of photoinduced hydrogen evolution systems using porphyrins are described above. To improve the hydrogen evolution yield, the following requirements are desired: 1) high efficiency of charge separation and 2) suppression of back electron transfer. Although a system with the above requirements has been investigated, an effective system has not yet been established. In the photoinduced hydrogen evolution with a four-component system, important steps in the system are the charge separation between photoexcited photosensitizer and electron carrier and suppression of back electron transfer. Effective charge separation and suppression of back electron transfer have been attempted using micellar systems,[33,34] colloidal inorganic suspension[35] or liposomes.[36] The micellar system of surfactant is especially useful for a homogeneous photoinduced hydrogen evolution.[37,38] The effects of various surfactants on photoinduced hydrogen evolution with water-soluble zinc porphyrins and hydrogenase are described. Fig. 3.36 shows the accumulation of reduced methyl viologen when a sample solution containing triethanolamine (as electron donor), ZnTPPS, methyl viologen and surfactant is irradiated. Remarkable increase is observed by the addition of an ionic surfactant such as CTAB or SDS. The accumulation of reduced methyl viologen strongly depends on the surfactant concentration and remarkable increase is observed above critical micellar concentration (CMC) of surfactant (Fig. 3.37).

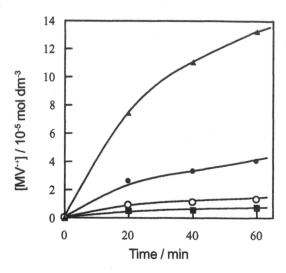

Fig. 3.36 Time dependence of reduced methyl viologen formation. A sample solution containing triethanolamine (0.25 mol dm⁻³), ZnTPPS (0.13 μmol dm⁻³) and methyl viologen (0.22 mmol dm⁻³) in 25 mmol dm⁻³ Tris-HCl buffer (pH=7.4) is irradiated at 30 °C. (■) without surfactant; (O) 40 mmol dm⁻³ Triton X-100 and (●) 25 mmol dm⁻³ CTAB. (Reproduced with permission from *J.Mol. Catal. A: Chem.*, **105** (1996) 125, Elsevier Science)

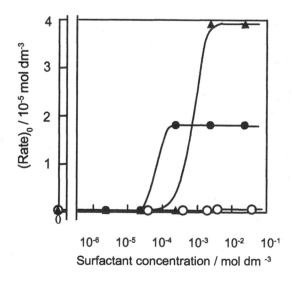

Fig. 3.37 Concentration effects of surfactant on the initial rate of methyl viologen photoreduction. ▲: SDS; ●: CTAB; O: Triton X-100.

The elementary processes of photoreduction of methyl viologen are studied using laser flash photolysis. The lifetimes of the photoexcited triplet state of ZnTPPS and reduced methyl viologen are important in the successive reaction. Typical decay curves of the photoexcited triplet state of ZnTPPS in the presence of surfactant micelles observed at 470 nm after laser flash are shown in Fig. 3.38. The lifetimes of the photoexcited triplet state of ZnTPPS in various surfactants are listed in Table 3.10. Although the lifetime increases dramatically in the presence of CTAB or Triton X-100, no increase in lifetime is observed in the case of SDS. Above CMC the lifetime of the photoexcited triplet state of ZnTPPS increases dramatically in the case of Triton X-100 or CTAB. Complexes are formed between ZnTPPS and surfactant by the addition of CTAB or Triton X-100. In the case of SDS, no complex is formed due to the electrostatic repulsion between the negatively charged ZnTPPS and the negative surface charge of SDS. The increase in lifetime is induced by the complex formation between a surfactant micelle such as CTAB or Triton X-100 and ZnTPPS.

Table 3.10 Lifetime of photoexcited triplet ZnTPPS (τ) in various surfactants

Surfactant	τ / ms
None	0.60
CTAB (25 mmol dm^{-3})	3.0
Triton X-100 (40 mmol dm^{-3})	4.2
SDS (25 mmol dm^{-3})	0.61

(Reproduced with permission from *J.Mol. Catal. A: Chem.*, **105** (1996) 125, Elsevier Science)

Figure 3.39 shows the decay curves of reduced methyl viologen in surfactant micelles observed at 605 nm, the absorption maximum of the reduced methyl viologen, after laser flash. Every decay followed second-order kinetics. The lifetimes of reduced methyl viologen and the back electron transfer rate constant (Table 3.11) show suppression of the back electron transfer between reduced methyl viologen and ZnTPPS cation radical in the presence of surfactant.

For CTAB and SDS, the retardation of back electron transfer is due to an electrostatic effect among the surface charge of CTAB or SDS micelle, ZnTPPS, and methyl viologen. The recombination

Table 3.11 Lifetime of reduced methyl viologen and rate constant for back electron transfer, k_b, in various surfactants

Surfactant	Lifetime of MV$^{+\cdot}$ / ms	k_b /mol^{-1} dm^3 s^{-1}
None	0.020	8.9×10^9
CTAB (25 mmol dm^{-3})	2.3	5.7×10^7
Triton X-100 (40 mmol dm^{-3})	1.0	1.6×10^8
SDS (25 mmol dm^{-3})	3.2	1.7×10^7

(Reproduced with permission from *J.Mol. Catal. A: Chem.*, **105** (1996) 125, Elsevier Science)

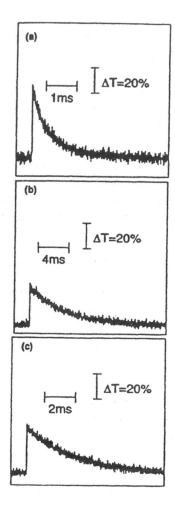

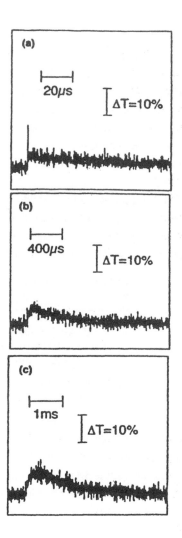

Fig. 3.38 Decay of photoexcited triplet state of ZnTPPS after laser flash monitored at 470 nm. A sample solution contains ZnTPPS (3.0×10^{-5} mol dm^{-3}) and surfactant in 25 mmol dm^{-3} Tris-HCl buffer (pH=7.4). (a) without surfactant; (b) 40 mmol dm^{-3} Triton X-100 and (c) 25 mmol dm^{-3} CTAB. (Reproduced with permission from *J.Mol. Catal. A: Chem.*, **105** (1996) 125, Elsevier Science)

Fig. 3.39 Decay of the reduced methyl viologen after laser flash monitored at 605 nm. A sample solution contains ZnTPPS (3.0×10^{-5} mol dm^{-3}), methyl viologen (0.75 mmol dm^{-3}), and surfactant in 25 mmol dm^{-3} Tris-HCl buffer (pH=7.4). (a) without surfactant; (b) 40 mmol dm^{-3} Triton X-100 and (c) 25 mmol dm^{-3} CTAB. (Reproduced with permission from *J.Mol. Catal. A: Chem.*, **105** (1996) 125, Elsevier Science)

of the reduced methyl viologen and ZnTPPS cation radical after the charge separation is suppressed by strong electrostatic repulsion induced by the adsorption of negatively charged ZnTPPS on the CTAB cationic micellar surface or positively charged methyl viologen on the SDS anionic micellar surface. On the other hand, for Triton X-100, the charge recombination of the reduced methyl viologen and ZnTPPS cation radical after the separation may be suppressed by complex formation of ZnTPPS and Triton X-100 by hydrophobic effect.

To establish an effective photoinduced hydrogen evolution system containing ZnTPPS and methyl viologen, effective photoreduction of methyl viologen is required. An effective system is achieved by the suppression of step 4 in Scheme 3-11. The mechanism of the methyl viologen photoreduction in the presence of surfactant is explained by the results of laser flash photolysis as shown in Scheme 3-11. Although the quenching reaction by methyl viologen is suppressed by the addition of surfactant (steps 2 and 3), little change in quenching efficiency is observed. On the other hand, by the addition of surfactant the back electron transfer between reduced methyl viologen and ZnTPPS cation radical (step 4) is suppressed 100 times as long as back electron transfer without surfactant. These results indicate that the effective photoreduction of methyl viologen is achieved more by the suppression of step 4 than by suppression of step 2 and step 3.

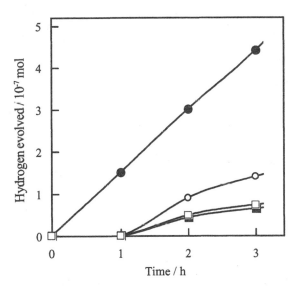

Fig. 3.40 Time dependence of hydrogen evolution. A sample solution containing triethanolamine (0.25 mol dm^{-3}), ZnTPPS (0.13 mmol dm^{-3}), methyl viologen (0.22 mmol dm^{-3}) and hydrogenase (50 ml) in 25 mmol dm^{-3} Tris-HCl buffer (pH=7.4) is irradiated at 30 $^{\circ}$C. (■) without surfactant; (O) 40 mmol dm^{-3} Triton X-100, (□) 25 mmol dm^{-3} SDS and (●) 25 mmol dm^{-3} CTAB. (Reproduced with permission from *J.Mol. Catal. A: Chem.*, **105** (1996) 125, Elsevier Science)

When a sample solution containing ZnTPPS, methyl viologen, triethanolamine, hydrogenase and surfactant is irradiated, hydrogen evolution is observed as shown in Fig. 3.40. The rate of hydrogen evolution in the presence of CTAB is about seven times as large as that in the absence of surfactant, and the rate of evolution in the presence of Triton X-100 is about twice as large as that in the absence of surfactant. For SDS, the effective photoreduction of methyl viologen is established. However, little hydrogen evolution is observed in the presence of SDS. On the other hand, in the presence of CTAB or Triton X-100, increase in hydrogen evolution is observed because of effective electron transfer from reduced methyl viologen to hydrogenase.

When a sample solution containing 15 mmol dm^{-3} ZnTPPS, 0.10 mmol dm^{-3} methyl viologen, 0.5 mol dm^{-3} triethanolamine, 25 mmol dm^{-3} CTAB and 0.35 unit hydrogenase in 25 mmol dm^{-3} of Tris-HCl (pH, 7.4) is irradiated at 30°C, effective photoinduced hydrogen evolution is observed as shown in Fig. 3.41. The rate of hydrogen evolved in the above conditions is estimated to be 6.4 x 10^{-7} mol h^{-1} i.e., 50 times as fast as that achieved under conventional conditions.[39] The high-yield photoinduced hydrogen evolution system is established by optimizing the reaction conditions (Φ=0.03).

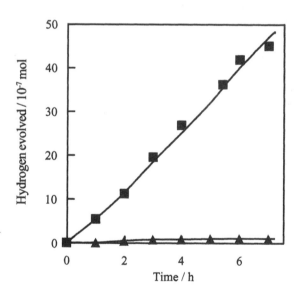

Fig. 3.41 Time dependence of hydrogen evolution. All experiments are carried out at 30°C. (■) triethanolamine (0.50 mol dm^{-3}), ZnTPPS (15 mmol dm^{-3}), methyl viologen (0.10 mmol dm^{-3}), CTAB (25 mmol dm^{-3}) and hydrogenase (50 ml) in 25 mmol dm^{-3} Tris-HCl buffer (pH=7.4). (▲) absence of CTAB. (Reproduced with permission from *Porphyrins*, **4** (1995) 129)

3.4 Intramolecular Electron Transfer in Acceptor-linked Porphyrins and Phthalocyanines

Recently, the initial photosynthesis mechanism has been elucidated by X-ray analysis of the complex of the reaction center in photosynthesis bacteria. The prosthetic group of the photosynthetic reaction center from *Rhodopseudomonas viridis* from X-ray crystalline analysis is shown in Fig. 3.42.[40] Light energy is absorbed by a chromophore near the reaction center and is transferred to a bacteriochlorophyll dimer. The electron from the excited bacteriochlorophyll transfers to bacteriopheofitin then to quinone

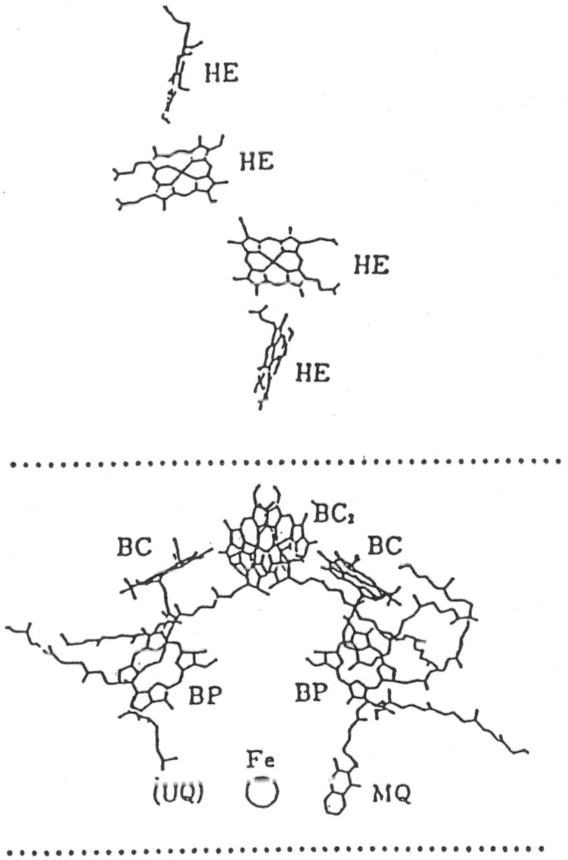

Fig. 3.42 Prosthetic group of photosynthetic reaction center from *Rhodopseudomonas viridis*. HE: Hemes, BC: Bacteriochlorophylls, BP: Bacteriopheophytins, MQ: Menaquinone, UQ: Ubiquinone, Fe: Non-heme iron. (Reproduced with permission by J. Deisserhofer et al., *Nature*, **318** (1985) 618, Macmillan Magazines Limited)

with lower energy. The charge-separated species with a long lifetime is achieved by suppression of back electron transfer inducing a long distance between the quinone anion radical and bacteriochlorophyll dimer cation radical. This charge-separated species is used as the reducing reagent and oxidizing reagent in the dark reaction.

The electron transfer processes in the photosynthesis reaction center occur via the photoexcited singlet state of chlorophyll in the region of picoseconds, and the quantum yield is *ca.* 100%. This effective photoexcited electron transfer occurs by the fixation of each chromophore of the photosynthesis reaction center.

To elucidate the mechanism of the photosynthesis reaction center and to establish effective artificial photosynthesis, many electron acceptor-linked porphyrins have been synthesized and their photochemical properties characterized. In this section, the photochemical properties of typical electron acceptor-linked porphyrins, quinone-linked porphyrins,[41-48] pyromellitimide-linked porphyrins,[49-52] fullerene (C_{60})-linked porphyrins [53-60] and viologen-linked porphyrins [61-70] are described.

3.4.1 Quinone-linked porphyrins[41-48]

Quinone-linked porphyrins are synthesized as model compounds of the photosynthesis reaction center and their photoexcited intramolecular electron transfer processes have been reported. The molecular structures of quinone-linked porphyrins are shown in Fig. 3.43. The distances between quinone and porphyrin of compound (**1**) and compound (**2**) are 0.82 and 0.86 nm, respectively, and the dihedral angle between the porphyrin ring and quinone of (**1**) and (**2**) is estimated to be 150° and 90°, respectively. From the fluorescence lifetime, the photoexcited intramolecular electron transfer rates of (**1**) and (**2**) are 3.3×10^8 s^{-1} and 6.5×10^7 s^{-1}, respectively. The conformation of (**1**) is more suitable for electron transfer compared with (**2**). Quinone-linked porphyrins ((**3**) and (**4**), isomers of (**1**) and (**2**)) are also prepared. From the fluorescence lifetime, the photoexcited intramolecular electron transfer rates of (**3**) and (**4**) are 3.5×10^9 s^{-1} and 4.5×10^8 s^{-1}, respectively. In these compounds, the photoexcited electron transfer rates depend on their conformation.

In quinone-linked porphyrins, although rapid photoexcited electron transfer occurs, the lifetime of charge-separated species is very short because of rapid charge recombination to ground state. To suppress the charge recombination and to achieve the formation of stable charge-separated species with a long lifetime, the quinone-linked porphyrins in which a different electron donor group or electron carrier group is connected are synthesized and their properties characterized.

Quinone-linked porphyrins with carotenoid are synthesized to attain the formation of stable charge-separated species with a long lifetime. The carotenoid chain serves as the electron donor in these molecules. The photoexcited electron transfer processes are shown in Scheme 3-12. In the case of compound (**5**), the lifetime of the final charge-separated species is *ca.* 300 ns in dichloromethane and *ca.* 2 μs in butylonitorile. The quantum yield for the formation of final charge-separated species is estimated to be 0.04. In compound (**6**), (methylene chain length is 2) the quantum yield is estimated to be 0.05, in compound (**7**) or (**8**), the yield is less than 0.04. To achieve the formation of charge-separated species with longer lifetimes, the porphyrin connected with two different quinones is synthesized

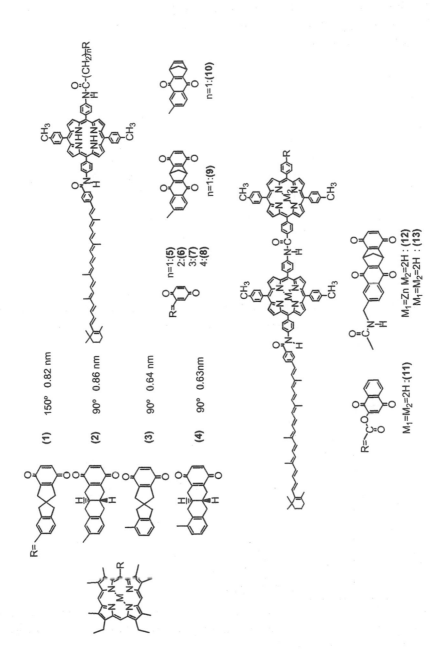

Fig. 3.43   The molecular structures of typical quinone-linked porphyrins.

and characterized. For compound (**9**), the photoexcited electron transfer processes are shown in
Scheme 3-13. The high quantum yield of the final charge-separated species is established to be 0.23
in dichloromethane and the lifetime is *ca.* 460 ns. On the other hand, in compound (**10**), naphtoquinone-
linked porphyrin, the quantum yield for the formation of a final charge-separated species is estimated to

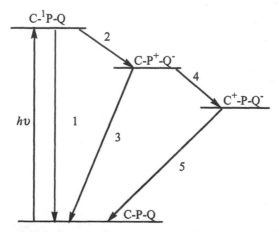

Scheme 3-12 Photoexcited electron transfer processes in quinone-carotenoid linked-porphyrin
(C-P-Q). (Reproduced with permission by D. Gust et al., *J. Am. Chem. Soc.*, **109**
(1987) 846, ACS)

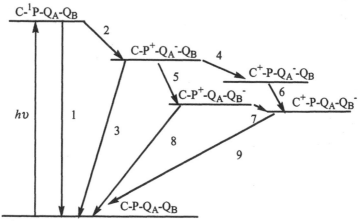

Scheme 3-13 Photoexcited electron transfer processes in bisquinone-carotenoid linked-porphyrin
(C-P-Q$_A$-Q$_B$). (Reproduced with permission by D. Gust et al., *J. Am. Chem. Soc.*,
110 (1988) 321, ACS)

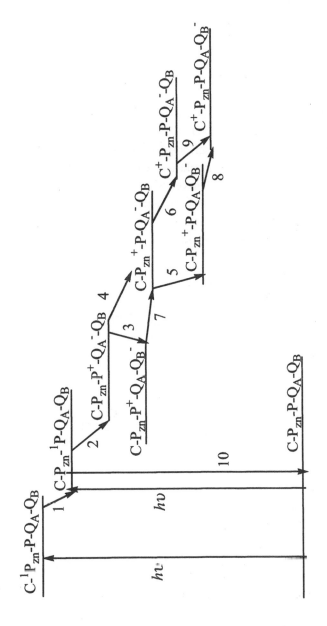

Scheme 3-14 Photoexcited electron transfer processes in bisquinone-carotenoid linked-diporphyrin (C-P$_{Zn}$-P-Q$_A$-Q$_B$). (Reproduced with permission by D. Gust et al., *J. Am. Chem. Soc.*, **110** (1988) 7567, ACS)

be 0.04 in dichloromethane and the lifetime is *ca.* 70 ns. Quinone-linked diporphyrin with carotenoid (**11-13**) are synthesized to achieve the formation of more stable charge-separated species. In these compounds, the photoexcited electron transfer processes are shown in Scheme 3-14. Especially in compound (**12**), the higher quantum yield of the final charge-separated species is 0.83 in chloroform and the lifetime is *ca.* 55 μs.

3.4.2 Pyromellitimide-linked porphyrins [49-52]

In quinone-linked porphyrins, the formation of charge-separated species with a long lifetime is attained by multi-step electron transfer using two different quinones. To form more stable charge-separated species with long lifetime, it is necessary to insert an effective electron carrier between the porphyrin and quinone. In this case, the electron carrier should have a redox potential between the porphyrin and quinone. Recently, pyromellitimide derivatives have been studied as effective electron carriers. Since the reduced pyromellitimide has the absorption band in the visible region, it is easy to monitor the reaction rate by the absorption of the reduced pyromellitimide. The molecular structures of pyromellitimide-linked porphyrins are shown in Fig. 3.44.

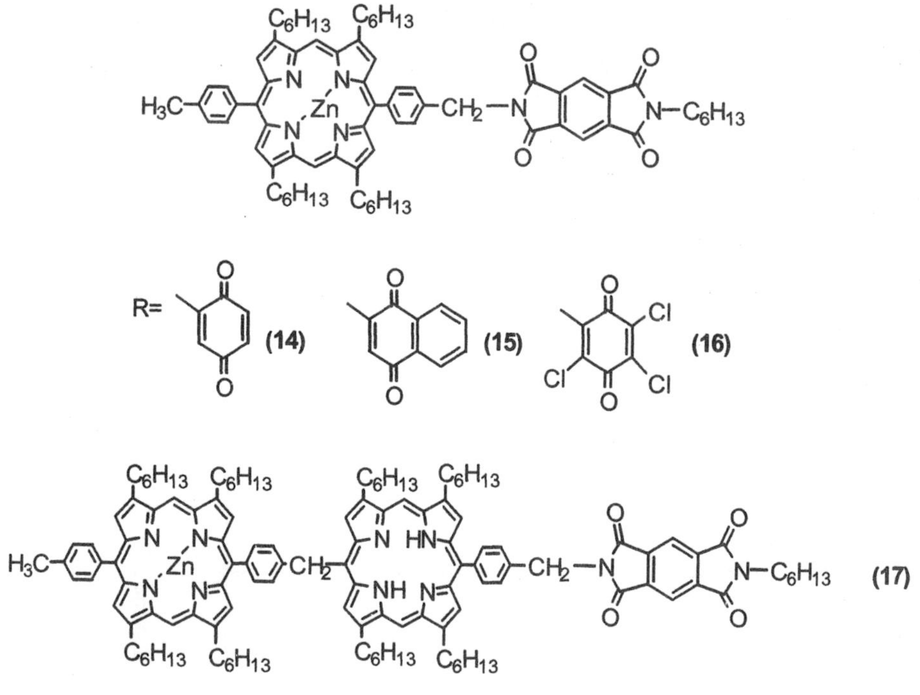

Fig. 3.44 The molecular structures of typical pyromellitimide-linked porphyrins. (Reproduced with permission by A. Osuka et al., *J. Am. Chem. Soc.*, **115** (1993) 4577, ACS)

The photoexcited electron transfer processes are shown in Scheme 3-15. The photoexcited intramolecular electron transfers are studied by picosecond laser flash photolysis. In all compounds, the reduced pyromellitimide is observed and the lifetime of the reduced pyromellitimide is estimated to be *ca.* 60 ps. An absorption band of 610 nm for zinc porphyrin cation radical is observed and the lifetime of the zinc porphyrin cation radical is in the region between 900 ps and 4 ns. As the lifetime of the zinc porphyrin cation radical is longer than that of reduced pyromellitimide, the electron transfers to the quinone via the reduced pyromellitimide followed by the formation of the charge-separated species. By using trichloro-1,3-quinone (**16**), the quantum yield of the formation of final charge-separated species is estimated to be 1.0, because the bonded chloride atom acts as an electron withdrawer and the electron transfers more easily. The quantum yield of the formation of the final charge-separated species is estimated to be 0.76 in compound (**14**). These results indicate that the effective electron transfer occurs via the bonded pyromellitimide.

In the compound (**17**), the photoexcited electron transfer processes are shown in Scheme 3-16. The quantum yield of the formation of the final charge-separated species is estimated to be 0.78 and the lifetime of the final charge-separated species is of the microsecond order. In this compound, the energy transfers the photoexcited singlet state of zinc porphyrin to zinc-free porphyrin and then the electron transfers from the photoexcited singlet state of zinc free porphyrin. The formation of the final charge-separated species from the photoexcited singlet state of zinc-free porphyrin is observed. The high efficient quantum yield of the formation of the final charge-separated species is caused by hole transfer from zinc-free porphyrin cation radical to zinc porphyrin. As mentioned above, the high quantum yield for formation of the final charge-separated species with longer lifetime is accomplished by hole transfer in pyromellitimide-linked porphyrins.

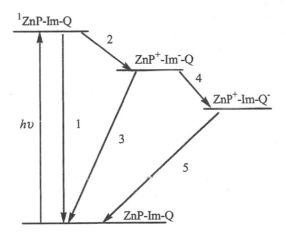

Scheme 3-15 Photoexcited electron transfer processes in pyromellitimid-quinone linked-porphyrin (ZnP-Im-Q). (Reproduced with permission by M. Ohkohchi et al., *J. Am. Chem. Soc.*, **115**, (1993) 12137, ACS)

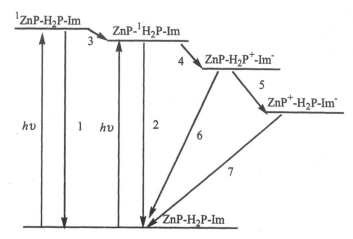

Schem 3-16 Photoexcited electron transfer processes in pyromellitimede-linked porphyrin (ZnP-H$_2$P-Im). (Reproduced with permission by A. Osuka et al., *Bull. Chem. Soc. Jpn.*, **68** (1995) 262)

3.4.3 Fullerene (C$_{60}$)-linked porphyrins [53-64]

Fullerene has unique electrochemical and photophysical properties both as an electron acceptor and a photosensitizer. Fullerene has moderate electron-accepting abilities similar to those of benzoquinones and naphthoquinones and reversibly accepts up to six electrons, and the energy levels of the first excited singlet state and triplet state are comparable to those of porphyrins and are much lower than those of acceptors such as quinones and pyromellitic diimides.[53-59] In this section, photochemical and photophysical properties of fullerene (C$_{60}$)-linked porphyrins are described (Fig. 3.45). Gust et al. have shown the synthesis and photochemical and photophysical properties of C$_{60}$-linked porphyrins (**18-19**).[60] C$_{60}$ are connected to porphyrin by a bicyclic rigid bridge. The photoexcited singlet lifetime of the porphyrin site in this compound is estimated to be 7 ps in toluene. These substantial quenchings are due to singlet-singlet energy transfer from the photoexcited singlet state of the porphyrin site to C$_{60}$. The lifetime of the singlet state of the C$_{60}$ moiety is estimated to be *ca.* 5 ps. On the other hand, the lifetime of the singlet state of C$_{60}$ model compound is 1.2 ns, indicating that the quenching is due to electron transfer to produce the charge-separated species (ZnP$^+$-C$_{60}^-$). The photoinduced electron transfer rate constant to yield ZnP$^+$-C$_{60}^-$ is *ca.* 2.0 x 10^{11} s^{-1}. The ultrafast intramolecular photochemical processes in these compounds can be explained as direct through-space interaction of the π-electron system, as opposed to through-bond interaction.

The UV-visible absorption spectra and redox potentials show that these compounds have a linear combination of the porphyrin and C$_{60}$ moieties, and there is no appreciable interaction between the two moieties. Fluorescence quenching of both excited singlet states of the porphyrin site by the

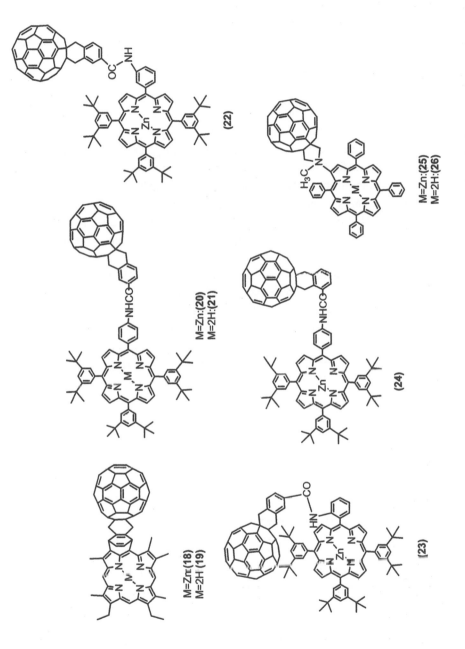

Fig. 3.45   The molecular structures of typical fullerene-linked porphyrins.

bonded C_{60} moieties is observed. Photoinduced electron transfer from the singlet state of the zinc porphyrin site to C_{60} is observed.

Related compounds with a variety of spacers between the porphyrin and C_{60} **(22-24)** have been synthesized.[61,62] Three different kinds of C_{60}-linked porphyrins have been synthesized by systematically changing the bonded position at the *meso*-phenyl ring. In THF the charge separation occurs from both the excited singlet state of the porphyrin and the C_{60} moieties, indicating that an increase in the absorption cross section by both chromophores results in efficient formation of the charge-separated state as shown in Fig. 3.46.[62] In benzene, on the other hand, the energy level of the charge-separated state is above that of ZnP-$^1C_{60}^*$ or at the same level. Therefore, the charge-separated state formed by the photoinduced electron transfer from the excited singlet state of porphyrin to the C_{60} energetically equilibrates with the locally excited singlet state of the C_{60}. Both the charge separation and charge recombination rates for compound **(23)** are much slower than those for the other C_{60}-linked porphyrins. Linkage dependence of the electron transfer rates can be explained reasonably by the following. Molecular orbital calculations on the linkage show that electron densities at *ortho*- and *para*-positions in HOMO and LUMO levels are much larger than the values at the *meta*-position. This indicates that electron coupling between the two chromophores through the *meta*-position is much smaller than that for the other two positions.

Reed et al. have reported a new type of porphyrin-C_{60} dyad, **(25)** and **(26)**, where the porphyrin and the C_{60} are linked via a pyrrolidine spacer.[63,64] In **(25)** and **(26)**, the π-electron systems of the two chromophores are separated with a center-to-center distance of 0.99 nm. In toluene, excited singlet states of the porphyrin decays in ~20 ps by singlet-singlet energy transfer to the C_{60}. The excited singlet state of C_{60} is unquenched and undergoes intersystem crossing to the triplet, which is in equilibrium with the porphyrin triplet state. In polar solvents such as benzonitrile, photoinduced electron transfer from the first excited singlet state of the porphyrin to the C_{60} competes with energy

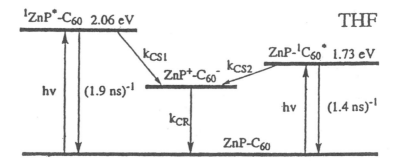

Fig. 3.46 Energy level diagram for the porphyrin and C_{60} first excited states of **20, 22-24** in THF. (Reproduced with permission by H. Imahori et al., *J. Am. Chem. Soc.*, **118** (1996) 1171, ACS)

transfer. On the other hand, the excited singlet state of the C_{60} is also quenched by electron transfer from the porphyrin. In total, the charge-separated state is produced with a quantum yield near unity. The resulting charge-separated state has a relatively long lifetime of 50 ps in (**25**) and 290 ps in (**26**).

3.4.4 Viologen-linked porphyrins [65-83]

The properties of quinone-linked porphyrins have been described in the previous section. Although the reduced quinone is biologically useful, it is difficult to establish a combination system of an enzyme and quinone for the decomposition of water, because the quinone does not have sufficient redox potential to decompose water. To establish a combination system consisting of an enzyme and acceptor-linked porphyrin, it is necessary to select a substrate for the enzyme as an electron acceptor.

Viologen derivatives can be the substrates for hydrogenase. In photoinduced hydrogen evolution with a four-component system of an electron donor, zinc porphyrin, an electron carrier and hydrogenase, the reaction occurs via the photoexcited triplet state instead of the photoexcited singlet state. Effective photoinduced hydrogen evolution via the photoexcited singlet state can be achieved using viologen-linked porphyrin.

Harriman et al. [65] have reported the synthesis of viologen-linked zinc porphyrins (**27**) (Fig. 3.47) and the effects of the methylene chain length between zinc porphyrin and viologen and the binding position of viologen on the fluorescence intensity of zinc porphyrin. Fluorescence quenching is observed by photoexcited intramolecular electron transfer via the photoexcited singlet state. For compound **27** with different methylene chain length, the order of efficiency of fluorescence quenching is the *ortho*, *meta* and *para* position because of the close distance between zinc porphyrin and viologen in *ortho* compounds. As the methylene chain is long enough in compound **27**, no effect of the binding position of viologen on the fluorescence intensity of zinc porphyrin is observed, indicating that no conformation change occurs when the methylene chain is long enough.

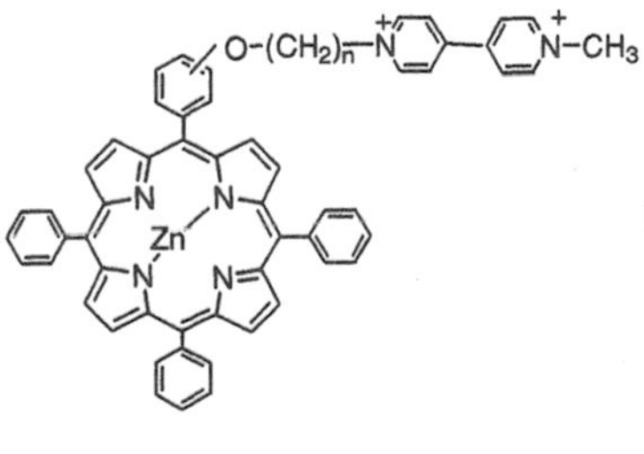

(27) 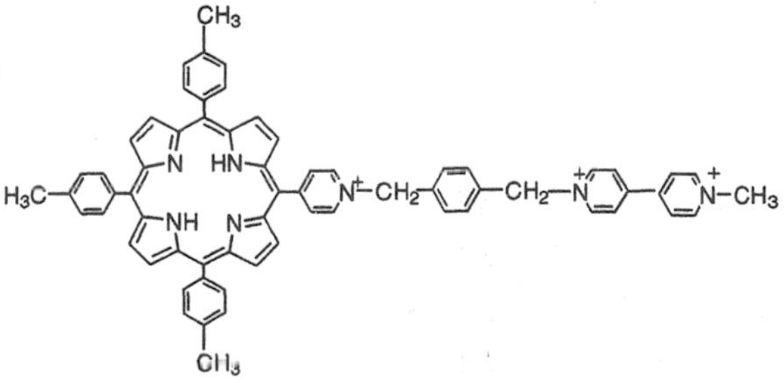   **(28)**

Fig. 3.47 The molecular structures of typical viologen-linked porphyrins.

Wrighton et al.[66] synthesized viologen-linked zinc porphyrins (**28**) (Fig. 3.47) and the photoexcited intramolecular electron transfer from the photoexcited porphyrin to viologen was studied by resonance Raman spectrum and fluorescence lifetimes. The lifetime of viologen-linked porphyrin and viologen-free porphyrin is estimated to be 5.3 and 9.0 ns, respectively, indicating that the fluorescence quenching is caused by the photoexcited intramolecular electron transfer via the photoexcited singlet state of the porphyrin. The photoexcited intramolecular electron transfer rate via the photoexcited singlet state of porphyrin is estimated to be $8.0 \times 10^7 \, s^{-1}$.

In the case of viologen-linked porphyrin (**29**) (Fig. 3.48), the effects of the methylene chain length between zinc porphyrin and viologen and the binding position of viologen on the photoexcited intramolecular electron transfer via the photoexcited singlet state of porphyrin have been reported. Table 3.12 shows the fluorescence intensities of p-PC_nV and m-PC_nV in acetonitrile and dimethylsulfoxide. The fluorescence of the porphyrin is quenched by the bonded viologen caused by

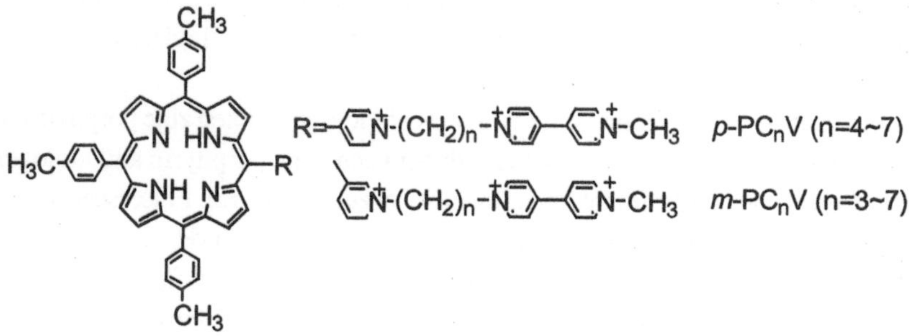

(**29**)

Fig. 3.48 The molecular structures of typical viologen-linked porphyrins. (Reproduced with permission from *J. Chem. Soc. Faraday Trans.* **1** (1990) 811, The Royal Society of Chemistry)

Table 3.12 Relative fluorescence intensities of p-PC_nV and m-PC_nV

Compound	I/I_0 (in DMF)	I/I_0 (in acetonitrile)
p-PC_4	1	1
p-PC_4V	0.57	0.39
p-PC_5V	0.82	0.52
p-PC_6V	0.91	0.78
p-PC_7V	0.88	0.73
m-PC_1	1	1
m-PC_3V	0.32	0.12
m-PC_4V	0.50	0.20
m-PC_5V	0.72	0.32
m-PC_6V	0.76	0.36

the photoexcited intramolecular electron transfer via the photoexcited singlet state. The rate constant for the photoexcited intramolecular electron transfer via the photoexcited singlet state of porphyrin is estimated to be 6.0×10^7 - 7.0×10^8 s^{-1} by the fluorescence lifetime.

　　Conformation of viologen-linked porphyrins have been studied by ^{1}H-NMR to gain insight into the relative geometry of the porphyrin and viologen during the electron transfer processes.[75] The orientation of the viologen with respect to the porphyrins is interpreted in terms of the rate of the intramolecular electron transfer between the two π-systems. ^{1}H-NMR spectra of viologen-linked porphyrins have been unambiguously assigned from the combined analyses of the connectivities in ^{1}H-^{1}H chemical shift correlated and nuclear Overhauser effect correlated spectroscopies. Chemical shifts for the viologen proton resonances of *p*- and *m*-PC$_n$V are given in Table 3.13. Pyridyl 2', 6' and 3', 5' protons of *p*-PC$_n$V resonate at 9.09 and 9.52 ppm, respectively, and the shifts of the other proton resonances are essentially identical to those of the corresponding resonances of *m*-PC$_n$V. These shifts are essentially independent of the sample concentration, indicating the absence of significant intramolecular interaction. The porphyrin ring current inducing shifts for the viologen proton resonances of *p*-PC$_n$V are calculated using the data obtained for appropriate model compounds. The ring current shifts (δ_{RC}) of the viologen proton resonances are summarized in Fig. 3.49 and their plots are individually compared with those of *m*-PC$_n$V in Fig. 3.50. The δ_{RC} value of the 1-CH$_2$ proton resonance for *p*-PC$_n$V is + 0.34 ppm and decreases to *ca.* 0 almost monotonically for protons farther away, along chemical bonds,

Table 3.13 Chemical shifts on viologen ^{1}H resonances of *p*-PC$_n$V (*n*=4-7) in DMSO-d_6

Protons	*p*-PC4V	*p*-PC5V	*p*-PC6V	*p*-PC7V
Methylene Chain				
1-CH$_2$	5.00 (+0.34)	4.95 (+0.34)	4.93 (+0.34)	4.92 (+0.34)
2-CH$_2$	2.35 (+0.34)	2.36 (+0.34)	2.30 (+0.37)	2.26 (+0.33)
3-CH$_2$	2.35 (+0.34)	1.64 (+0.28)	1.62 (+0.28)	1.58 (+0.24)
4-CH$_2$	4.92 (+0.31)	2.24 (+0.26)	1.62 (+0.25)	1.55 (+0.13)
5-CH$_2$		4.84 (+0.17)	2.14 (+0.21)	1.51 (+0.13)
6-CH$_2$			4.79 (+0.13)	2.08 (+0.15)
7-CH$_2$				4.75 (+0.08)
Bipyridyl				
α	9.50 (+0.17)	9.47 (+0.12)	9.45 (+0.11)	9.42 (+0.09)
β	8.85 (+0.07)	8.84 (+0.06)	8.83 (+0.06)	8.79 (+0.02)
γ	8.79 (+0.06)	8.70 (-0.04)	8.79 (+0.04)	8.72 (-0.01)
δ	9.31 (+0.03)	9.15 (-0.13)	9.28 (+0.01)	9.24 (-0.03)
N-Me	4.46 (+0.02)	4.28 (-0.16)	4.44 (0.00)	4.41 (-0.02)

(Reproduced with permission from *Bull. Chem. Soc. Jpn.*, **64** (1991) 1392)

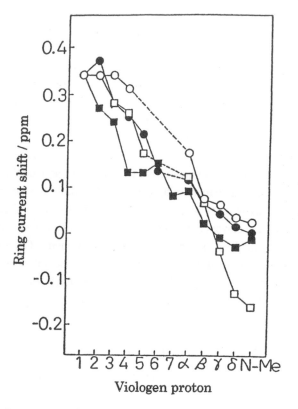

Fig. 3.49 Plots of the ring current-induced chemical shift against viologen protons of p-PC$_n$V in DMSO-d$_6$. O: n=4, □: n=5, ●: n=6, ■: n=7. (Reproduced with permission from *Bull. Chem. Soc. Jpn.*, **64** (1991) 1392)

from the porphyrin ring. Similarity in their δ_{RC} plots indicates that the orientation of the viologen with respect to the porphyrin is not significantly different among the p-PC$_n$V. These results suggest that the methylene chain of p-PC$_n$V is in an energetically stable all-*trans* extended conformation as found for m-PC$_n$V. Anomalously large negative δ_{RC} detected for the d and N-Me protons of p-PC$_5$V may not be explained in terms of its conformational features. Assuming the extended all-*trans* conformation for the methylene chain, the closest edge-to-edge distances between the two π-system (R$_{\pi\pi}$) in p-PC$_n$V with n=4, 5, and 6 are 1.03, 1.13, and 1.27 nm, respectively, and are compared with R$_{\pi\pi}$ of 0.85, 0.92, and 1.09 nm for m-PC$_n$V with the corresponding n value, respectively. Therefore, the fact that p-PC$_n$V (n>4) does not exhibit intramolecular electron transfer leads to the conclusion that the electron transfer in DMSO occurs when R$_{\pi\pi}$ < *ca.* 1.1 nm and that the electron is not transferred through the methylene chain. R$_{\pi\pi}$ of 1.1 nm is greater than the theoretical estimate, 0.8 nm, for electron transfer by a short-range tunneling mechanism.[76] Furthermore, R$_{\pi\pi}$ should be considered to be a semiquantitative

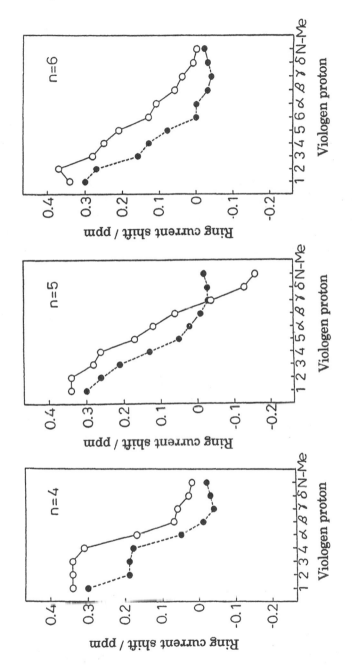

Fig. 3.50 Plot of the ring current-induced chemical shift against viologen protons of *p*-PC$_n$V in DMSO-d$_6$ (O). Similar data for *m*-PC$_n$V(●) are also shown for comparison. (Reproduced with permission from *Bull. Chem. Soc. Jpn.*, **64** (1991) 1392).

parameter for interpreting the intramolecular electron transfer rate in terms of the molecular structure of those viologen-linked porphyrins because it does not account for the fact that *m*-PC$_3$V and *p*-PC$_4$V exhibit similar rates. Considering both the conformational flexibility of the methylene chain in solution and the time scale, ns, of the process, R$_{\pi\pi}$ calculated for a molecular structure inferred from NMR data may not correlate simply to the electron transfer rate. A conformational change around one of the C-C bonds in the methylene chain, for example, *trans* to *gauche*, alters R$_{\pi\pi}$ by as much as *ca.* 10% and *ca.* 15% for *p*- and *m*-PC$_n$V, respectively.

Shafirovich et al. have published a series of papers dealing with multistep photoinduced electron transfers in systems consisting of photoexcited porphyrin donors and viologen and quinone acceptors (Fig. 3.51) (**30**).[77] The porphyrin-viologen dyad displays interesting magnetic field effects following electron transfer from the porphyrin excited triplet state to viologen. Formation of the radical ion pair within a magnetic field increases its lifetime by a factor of 11. From the fluorescence decay properties, the molecule is primarily in two conformations, one in which the porphyrin and viologen are folded and the other is extended. Water-soluble viologen-linked zinc anionic porphyrins with a carboxyl group (Fig. 3.52) (**31**) have been synthesized and their photochemical properties investigated by fluorescence emission spectra, fluorescence lifetime and laser flash photolysis. Table 3.14 shows the fluorescence intensities of ZnP(CO$_2$)$_3$C$_n$V in water. The fluorescence of porphyrin is quenched by the bonded

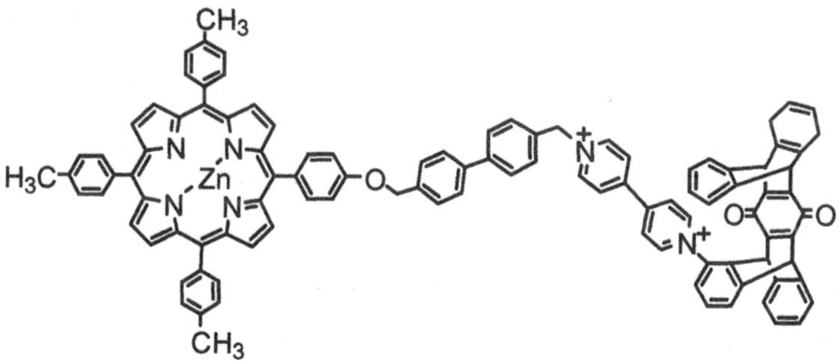

Fig. 3.51 The molecular structure of quinone-viologen-linked porphyrin.

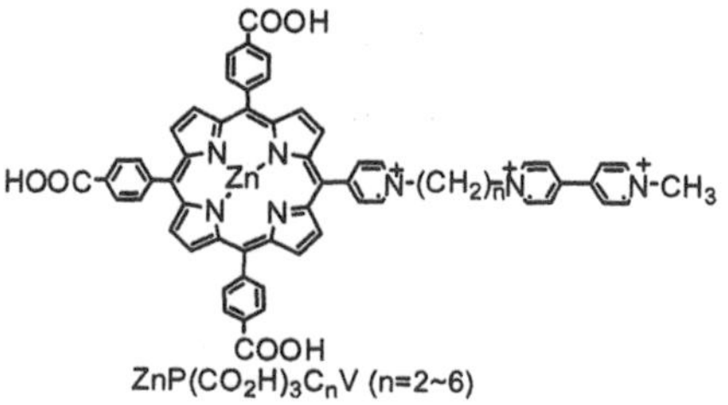

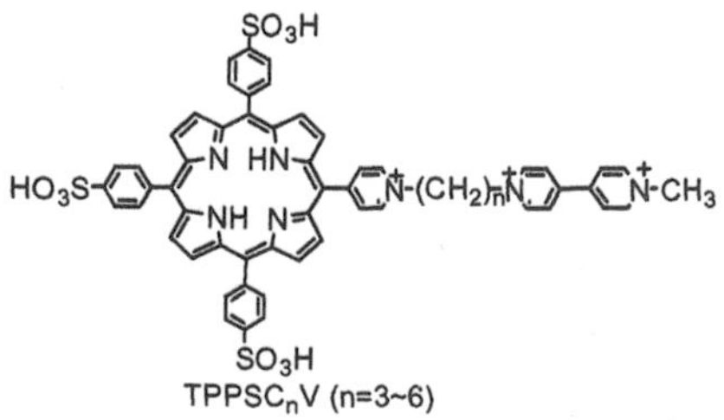

Fig. 3.52 The molecular structures of typical water-soluble viologen-linked anionic porphyrins.

viologen. Table 3.15 shows the fluorescence lifetimes of $ZnP(CO_2)_3C_nV$ in water. The lifetime of $ZnP(CO_2)_3C_nV$ is shorter than that of viologen-free zinc porphyrin, indicating that the photoexcited singlet state of porphyrin is quenched by the bonded viologen. Table 3.16 shows the intramolecular electron transfer rate from the photoexcited singlet state of zinc porphyrin site to the bonded viologen. Table 3.17 shows the lifetimes of the photoexcited triplet state of $ZnP(CO_2)_3C_nV$. The lifetime of

Table 3.14 Relative fluorescence intensities of $ZnP(CO_2)_3CnV$

Compounds	I / I_0
$ZnP(CO_2)_3Me$	1
$ZnP(CO_2)_3C_2V$	0.56
$ZnP(CO_2)_3C_3V$	0.56
$ZnP(CO_2)_3C_4V$	0.52
$ZnP(CO_2)_3C_5V$	0.77
$ZnP(CO_2)_3C_6V$	0.76

(Reproduced with permission from *Bull. Chem. Soc. Jpn.*, **63** (1990) 2922)

Table 3.15 Fluoresence lifetimes of $ZnP(CO_2)_3C_nV$

	τ_s / ns	A_s / %	τ_l / ns	A_l / %
$Zn\text{-}PC_3(CH_3)$			6.76	100
$Zn\text{-}PC_3(C_2V)$	2.01	54.8	7.52	45.2
$Zn\text{-}PC_3(C_3V)$	1.50	68.8	7.47	31.2
$Zn\text{-}PC_3(C_4V)$	2.05	60.7	7.93	39.3
$Zn\text{-}PC_3(C_5V)$	2.17	64.0	8.09	36.0
$Zn\text{-}PC_3(C_6V)$	1.88	74.4	8.20	25.6

(Reproduced with permission from *Bull. Chem. Soc. Jpn.*, **63** (1990) 2922)

Table 3.16 Intramolecular electron transfer rate constant via the photoexcited singlet state, k_{et}, in $Zn\text{-}PC_3(C_nV)$

Compound	k_{et} /s^{-1}
$Zn\text{-}PC_3(C_2V)$	3.7 x 10^8
$Zn\text{-}PC_3(C_3V)$	5.3 x 10^8
$Zn\text{-}PC_3(C_4V)$	3.7 x 10^8
$Zn\text{-}PC_3(C_5V)$	3.4 x 10^8
$Zn\text{-}PC_3(C_6V)$	4.1 x 10^8

(Reproduced with permission from *Bull. Chem. Soc. Jpn.*,**63** (1990) 2922)

Table 3.17 Lifetimes of excited triplet state of Zn-PC$_3$(C$_n$Br) and Zn-PC$_3$(C$_n$V)

	Lifetime / 10^{-6} s		Lifetime / 10^{-6} s
Zn-PC$_3$(C$_2$Br)	237	Zn-PC$_3$(C$_2$V)	200
Zn-PC$_3$(C$_3$Br)	295	Zn-PC$_3$(C$_3$V)	286
Zn-PC$_3$(C$_4$Br)	290	Zn-PC$_3$(C$_4$V)	203
Zn-PC$_3$(C$_5$Br)	205	Zn-PC$_3$(C$_5$V)	322
Zn-PC$_3$(C$_6$Br)	240	Zn-PC$_3$(C$_6$V)	200

(Reproduced with permission from *Bull. Chem. Soc. Jpn.,* **63** (1990) 2922)

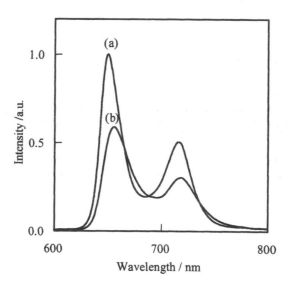

Fig. 3.53 Fluorescence spectra of TPPS (a) and TPPSC$_3$V (b) in water containing 1% Triton X-100. The excitation wavelength is 421 nm. (Reproduced with permission from *J. Mol. Catal. A: Chem.,* **126** (1997) 13, Elsevier Science)

ZnP(CO$_2$)$_3$C$_n$V is almost the same as that of viologen-free zinc porphyrin, indicating that the photoexcited triplet state is not quenched. These results indicate that the intramolecular electron transfer between the photoexcited porphyrin to viologen occurs via the photoexcited singlet state.

Water-soluble viologen-linked anionic porphyrins with a sulfo-group (TPPSC$_n$V) (**33**) have been synthesized and their photochemical properties investigated.[78,79] The fluorescence spectra of TPPSC$_3$V and TPPS are shown in Fig. 3.53. The shape of the fluorescence spectrum of TPPSC$_3$V is the same as that of TPPS, but the fluorescence intensity of TPPSC$_3$V is lower than that of TPPS. Table 3.18 shows the fluorescence intensities of TPPSC$_n$V in water. The fluorescence of porphyrin is quenched by the bonded viologen. The fluorescence decays of TPPSC$_3$V and TPPS are shown in Fig. 3.54. Fig. 3.55 shows the decay of the photoexcited triplet state of TPPSC$_3$V and TPPS after laser irradiation

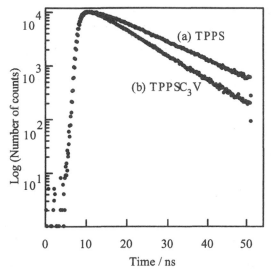

Fig. 3.54 Fluorescence decay curve of TPPS (a) and TPPSC$_3$V (b) in water containing 1% Triton X-100. The excitation wavelength is 350 nm. (Reproduced with permission from *J. Mol. Catal. A:Chem.,***126** (1997) 13 , Elsevier Science)

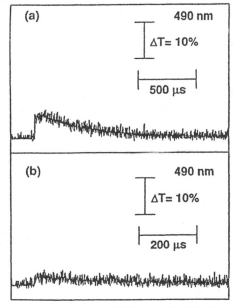

Fig. 3.55 Decay of the photoexcited triplet state of TPPS (a) and TPPSC$_3$V (b) in water containing 1% Triton X-100 monitored at 470 nm. The excitation wavelength is 532 nm. (Reproduced with permission from *J. Mol. Catal. A:Chem.,***126** (1997) 13, Elsevier Science)

with 532 nm monitored at 470 nm. Table 3.19 shows the lifetimes of the photoexcited triplet state of TPPSC$_n$V. The lifetime of TPPSC$_n$V is almost the same as that of TPPS, indicating that the photoexcited triplet state is not quenched. These results indicate that the intramolecular electron transfer between the photoexcited porphyrin site to viologen occurs via the photoexcited singlet state.

To accomplish the formation of charge-separated species with long lifetime, two different viologens are connected by zinc porphyrin; bisviologen-linked zinc porphyrin (**33**) (Fig. 3.56) has been

Table 3.18 Relative fluorescence intensities of TPPS and TPPSC$_n$V

Compound	I/I$_0$
TPPS	1
TPPSC$_3$V	0.58
TPPSC$_4$V	0.70
TPPSC$_5$V	0.56
TPPSC$_6$V	0.62

Excitation wavelength : 421 nm

(Reproduced with permission from *J. Mol. Catal. A: Chem.*, **126** (1997) 13, Elsevier Science)

Table 3.19 Fluorescence lifetimes of TPPS and TPPSC$_n$V

Compound	τ_{flu}/ ns
TPPS	12.7
TPPSC$_3$V	9.10
TPPSC$_4$V	10.4
TPPSC$_5$V	10.0
TPPSC$_6$V	10.1

Excitation wavelength : 350 nm

(Reproduced with permission from *J. Mol. Catal. A:Chem.*, **126** (1997) 13, Elsevier Science)

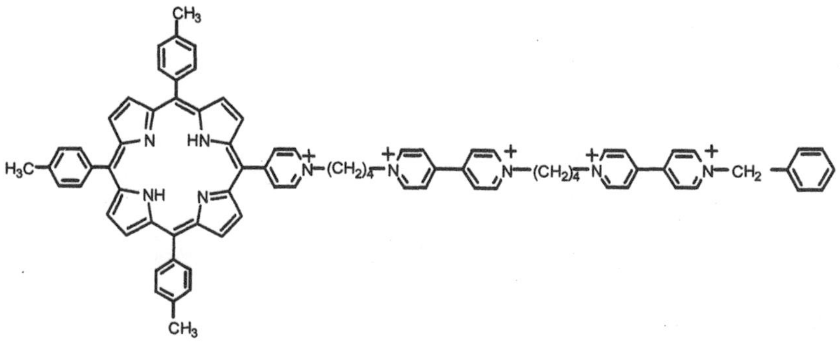

(33)

Fig. 3.56 The molecular structure of bisviologen-linked porphyrin.

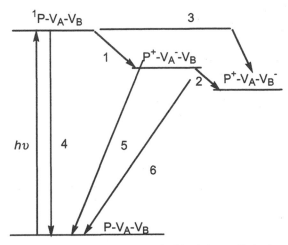

Scheme 3-17   Photoexcited electron transfer process in bisviologen-linked porphyrin (P-V_A-V_B). (Reproduced with permission by M. Calvin, D.R. Kearns, *US Patent,* 3,057,947 CA (1963) 2990a)

synthesized and characterized. In this compound, the photoexcited electron transfer processes are shown in Scheme 3-17. The fluorescence of porphyrin is quenched by the bonded viologen caused by the photoexcited intramolecular electron transfer via the photoexcited singlet state. The photoexcited singlet state of zinc porphyrin is quenched by the bonded viologen, but the photoexcited triplet state is not quenched according to the fluorescence lifetime and laser flash photolysis.

Water-soluble viologen-linked and bisviologen-linked cationic zinc porphyrins (**34**) (Fig. 3.57)

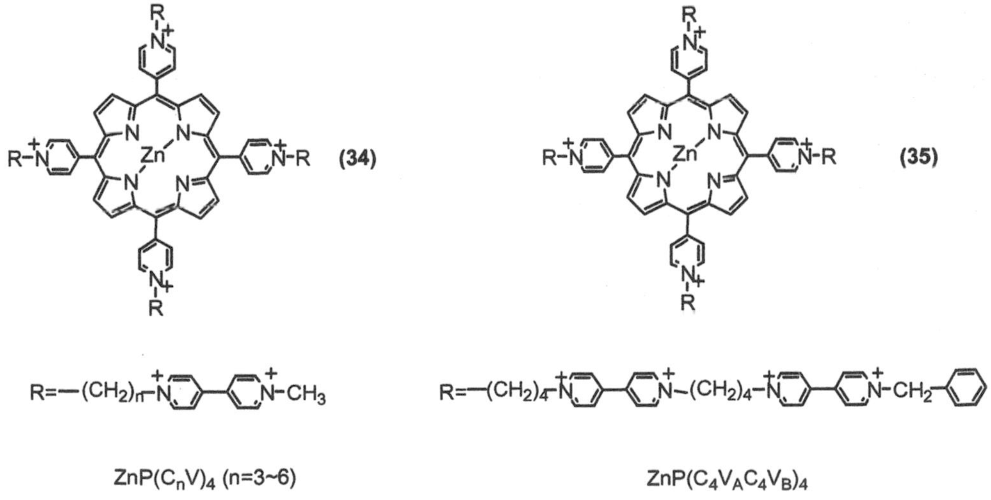

Fig. 3.57   The molecular structures of typical water-soluble viologen-linked cationic porphyrins.

$ZnP(C_nV)_4$ have been synthesized and their photochemical properties investigated.[80,81] The fluorescence spectra of $ZnP(C_3V)_4$ and viologen-free zinc-porphyrin, $ZnP(C_3)_4$ are shown in Fig. 3.58. The shape of the fluorescence spectrum of $ZnP(C_3V)_4$ is the same as that of $ZnP(C_3)_4$, but the fluorescence intensity of $ZnP(C_3V)_4$ is lower than that of $ZnP(C_3)_4$. Table 3.20 shows the fluorescence intensities of $ZnP(C_nV)_4$ in water. The fluorescence of porphyrin is quenched by the bonded viologen caused by the photoexcited intramolecular electron transfer via the photoexcited singlet state. The fluorescence decays of $ZnP(C_3V)_4$ and $ZnP(C_3)_4$ are shown in Fig. 3.59. Table 3.21 shows the fluorescence lifetime of $ZnP(C_nV)_4$ in water. The fluorescence decay of $ZnP(C_nV)_4$ consists of two components (slower and faster lifetimes) and the faster lifetime component makes a large contribution. The lifetimes of $ZnP(C_nV)_4$ are shorter than those of viologen free-zinc porphyrins, indicating that the photoexcited singlet state of zinc porphyrin is quenched by the bonded viologen. The intramolecular electron transfer rate constant from the

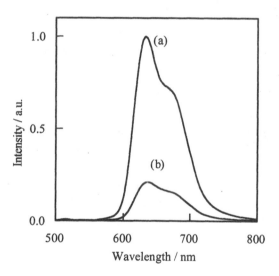

Fig. 3.58 Fluorescence spectra of $ZnP(C_3)_4$ (a) and $ZnP(C_3V)_4$ (b). The excitation wavelength is 438 nm. (Reproduced with permission from *J. Photochem. Photobiol. A: Chem.*, **98** (1996) 59, Elsevier Science)

Table 3.20 Relative fluorescence intensities of $ZnP(C_nV)_4$

Compound	I / I_0
$ZnP(C_3V)_4$	0.20
$ZnP(C_4V)_4$	0.65
$ZnP(C_5V)_4$	0.73
$ZnP(C_6V)_4$	0.60

I_0 is the fluorescence intensities of viologen free zinc porphyrin ($ZnP(C_n)_4$). Excitation wavelength was 438 nm.

(Reproduced with permission from *J. Porphyrins Phthalocyanines*, **2** (1998) 201, John Wiley & Sons Limited)

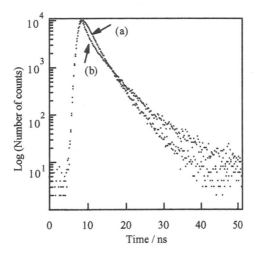

Fig. 3.59 Fluorescence decay curve of ZnP(C$_3$)$_4$ (a) and ZnP(C$_3$V)$_4$ (b). The excitation wavelength is 350 nm. The solid line indicates the lamp decay. (Reproduced with permission from *J. Porphyrins Phthalocyanines*, **2** (1998) 201, John Wiley & Sons Limited)

Table 3.21 Fluorescence lifetime of ZnP(C$_n$V)$_4$ and ZnP(C$_n$)$_4$

Compound	τ_s / ns (A$_1$ / %)	τ_l / ns (A$_2$ / %)
ZnP(C$_3$V)$_4$	0.24 (96.8)	4.4 (3.2)
ZnP(C$_3$)$_4$	1.9	
ZnP(C$_4$V)$_4$	0.90 (91.5)	3.9 (8.5)
ZnP(C$_4$)$_4$	1.7	
ZnP(C$_5$V)$_4$	0.90 (84.9)	3.0 (15.1)
ZnP(C$_5$)$_4$	1.8	
ZnP(C$_6$V)$_4$	0.90 (84.5)	2.7 (15.5)
ZnP(C$_6$)$_4$	1.8	

Excitation wavelength : 350 nm

(Reproduced with permission from *J. Porphyrins Phthalocyanines*, **2** (1998) 201, John Wiley & Sons Limited)

photoexcited singlet state of zinc porphyrin site to the bonded viologen in ZnP(C$_n$V)$_4$ based on the fluorescence lifetime is shown in Table 3.22. The photoexcited singlet state is quenched by the bonded viologen by the measurement of the fluorescence lifetime. Fig. 3.60 shows the decay of the photoexcited triplet state of ZnP(C$_3$V)$_4$ and ZnP(C$_3$)$_4$ after laser irradiation with 532 nm monitored at 490 nm. Table 3.23 shows the lifetimes of the photoexcited triplet state of ZnP(C$_n$V)$_4$. The lifetime of ZnP(C$_n$V)$_4$ is shorter than that of ZnP(C$_n$)$_4$, indicating that the photoexcited triplet state is also quenched by the bonded viologen. These results indicate that the intramolecular electron transfer between the photoexcited porphyrin site to viologen occurs via the photoexcited singlet state and triplet state.

Table 3.22 Intramolecular electron transfer rate constant via the photoexcited
singlet state, k_{et}, in $ZnP(C_nV)_4$

Compound	$k_{et}\ /s^{-1}$
$ZnP(C_3V)_4$	3.9×10^9
$ZnP(C_4V)_4$	8.5×10^8
$ZnP(C_5V)_4$	7.8×10^8
$ZnP(C_6V)_4$	8.8×10^8

(Reproduced with permission from *J. Porphyrins Phthalocyanines,* **2** (1998) 201,
John Wiley & Sons Limited)

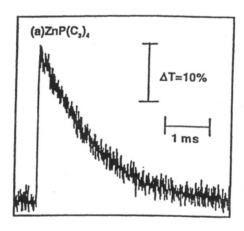

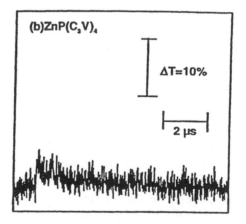

Fig. 3.60 Decay of the photoexcited triplet state of $ZnP(C_3)_4$. (a) and $ZnP(C_3V)_4$ (b) monitored at 490
nm. The excitation wavelength is 532 nm. (Reproduced with permission from *J. Porphyrins
Phthalocyanines,* **2** (1998) 201, John Wiley & Sons Limited)

Table 3.23 Lifetime of the photoexcited triplet state of ZnP(C$_n$V)$_4$ and ZnP(C$_n$)$_4$

Compound	τ_{trip} / µs
ZnP(C$_3$V)$_4$	1.5
ZnP(C$_3$)$_4$	1300
ZnP(C$_4$V)$_4$	3.4
ZnP(C$_4$)$_4$	1300
ZnP(C$_5$V)$_4$	4.2
ZnP(C$_5$)$_4$	1400
ZnP(C$_6$V)$_4$	4.2
ZnP(C$_6$)$_4$	1300

Excitation wavelength : 532 nm

(Reproduced with permission from *J. Porphyrins Phthalocyanines*, **2** (1998) 201, John Wiley & Sons Limited)

Water-soluble bisviologen-linked zinc porphyrin (ZnP(C$_4$V$_A$C$_4$V$_B$)$_4$) (Fig. 3.58) (**35**) has been synthesized and characterized.[82,83] The fluorescence spectra of ZnP(C$_4$V$_A$C$_4$V$_B$)$_4$ and viologen-free zinc-porphyrin, ZnP(C$_4$)$_4$ are shown in Fig. 3.61. The shape of the fluorescence spectrum of ZnP(C$_4$V$_A$C$_4$V$_B$)$_4$ is the same as that of ZnP(C$_4$)$_4$, but the fluorescence intensity of ZnP(C$_4$V$_A$C$_4$V$_B$)$_4$ is lower than that of ZnP(C$_4$)$_4$. The fluorescence of porphyrin is quenched by the bonded viologen caused by the photoexcited intramolecular electron transfer via the photoexcited singlet state. The fluorescence decays of ZnP(C$_4$V$_A$C$_4$V$_B$)$_4$ and ZnP(C$_4$)$_4$ are shown in Fig. 3.62. The fluorescence decay of ZnP(C$_4$V$_A$C$_4$V$_B$)$_4$ consists of two components (slow and fast lifetimes) and the faster lifetime component makes a large contribution. The lifetimes of ZnP(C$_4$V$_A$C$_4$V$_B$)$_4$ are shorter than those of

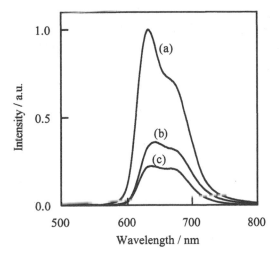

Fig. 3.61 Fluorescence spectra of ZnP(C$_4$)$_4$ (a), ZnP(C$_4$VC$_4$)$_4$ (b) and ZnP(C$_4$V$_A$C$_4$V$_B$)$_4$ (c). The excitation wavelength is 438 nm. (Reproduced with permission from *Inorg. Chim. Acta,* **267** (1998) 257, Elsevier Science)

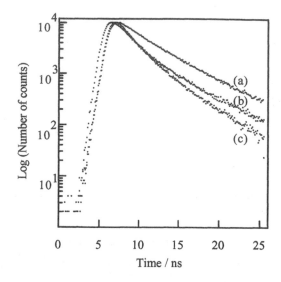

Fig. 3.62 Fluorescence decay curve of ZnP(C$_4$)$_4$ (a), ZnP(C$_4$VC$_4$)$_4$ (b) and ZnP(C$_4$V$_A$C$_4$V$_B$)$_4$ (c). The excitation wavelength is 350 nm. The solid line indicates the lamp decay. (Reproduced with permission from *J. Mol. Catal. A: Chem.*, **120** (1997) 5, Elsevier Science)

ZnP(C$_4$)$_4$, indicating that the photoexcited singlet state of zinc porphyrin is quenched by the bonded viologen. The intramolecular electron transfer rate constant from the photoexcited singlet state of zinc porphyrin site to the bonded viologen in ZnP(C$_4$V$_A$C$_4$V$_B$)$_4$ based on the fluorescence lifetime is shown in Table 3.24. The photoexcited singlet state is quenched by the bonded viologen by the measurement of the fluorescence lifetime. Fig. 3.63 shows the decay of the photoexcited triplet state of ZnP(C$_4$V$_A$C$_4$V$_B$)$_4$ after laser irradiation with 532 nm monitored at 490 nm. Table 3.25 shows the lifetimes of the photoexcited triplet state of ZnP(C$_4$V$_A$C$_4$V$_B$)$_4$ and ZnP(C$_4$)$_4$. The lifetime of ZnP(C$_4$V$_A$C$_4$V$_B$)$_4$ is shorter than that of ZnP(C$_4$)$_4$, indicating that the photoexcited triplet state is not quenched.

Table 3.24 Intramolecular electron transfer rate constant via the photoexcited singlet state, k_{et}, in ZnP(C$_4$V$_A$C$_4$V$_B$)$_4$ and ZnP(C$_4$VC$_4$)$_4$

Compound	k_{et}/ s^{-1}
ZnP(C$_4$V$_A$C$_4$V$_B$)$_4$	5.0 x 10^8
ZnP(C$_4$VC$_4$)$_4$	8.5 x 10^8

(Reproduced with permission from *J. Mol. Catal. A: Chem.*, **120** (1997) 5, Elsevier Science)

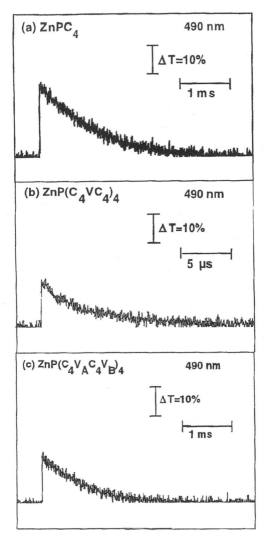

Fig. 3.63 Decay of the photoexcited triplet state of $ZnP(C_4)_4$ (a), $ZnP(C_4VC_4)_4$ (b) and $ZnP(C_4V_AC_4V_B)_4$ (c) monitored at 490 nm. The excitation wavelength is 532 nm. (Reproduced with permission from *J. Mol. Catal. A: Chem.*, **120** (1997) 5, Elsevier Science)

Table 3.25 Lifetime of the photoexcited triplet state of $ZnP(C_4V_AC_4V_B)_4$ and $ZnP(C_4VC_4)_4$ and $ZnP(C_4)_4$

Compound	τ_{trip} / µs
$ZnP(C_4V_AC_4V_B)_4$	890
$ZnP(C_4VC_4)_4$	9.5
$ZnP(C_4)_4$	1300

Excitation wavelength : 532 nm
(Reproduced with permission from *J. Mol. Catal. A: Chem.*, **120** (1997) 5, Elsevier Science)

3.4.5 Acceptor-linked phthalocyanines

Liu et al. utilized a soluble phthalocyanine derivative to produce viologen-linked phthalocyanine (Fig. 3.64) (**36**).[84] These workers have reported indications of electron transfer via the photoexcited singlet state and triplet state. In the former case, fluorescence quenching is observed and evidence for both static and dynamic quenching is presented. This indicates that conformational heterogeneity similar to that indicated by Levin et al. is also present in these systems. On a microsecond time scale there is evidence of the formation of a ZnPc$^+$-V$^+$ ion pair. Additional fluorescence quenching in the triad molecule may be due to enhanced nonradiative decay influenced by electronic coupling between the two macrocycles.

Li et al. have extended this approach to produce a porphyrin-phthalocyanine-quinone triad molecule (Fig. 3.64) (**37**).[85] For this compound, ZnPc is easier to oxidize than ZnTPP by *ca.* 0.17 V. Thus, the porphyrin functions only as an antenna molecule.

As described above, many studies on viologen-linked porphyrins have been carried out. Since the photoexcited state of zinc porphyrin is quenched by the bonded viologen, the establishment of an effective photoinduced hydrogen evolution system can be anticipated using these compounds.

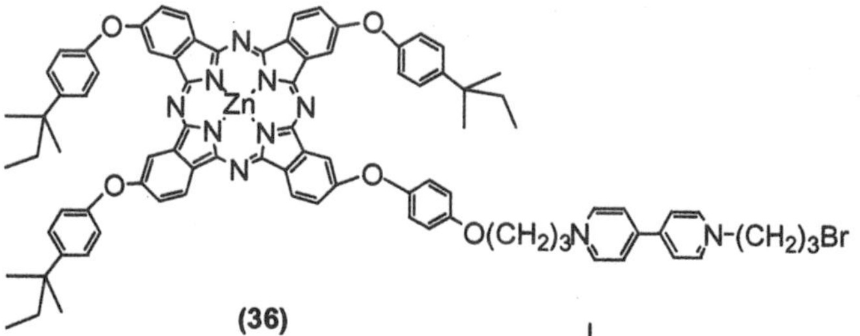

(36)

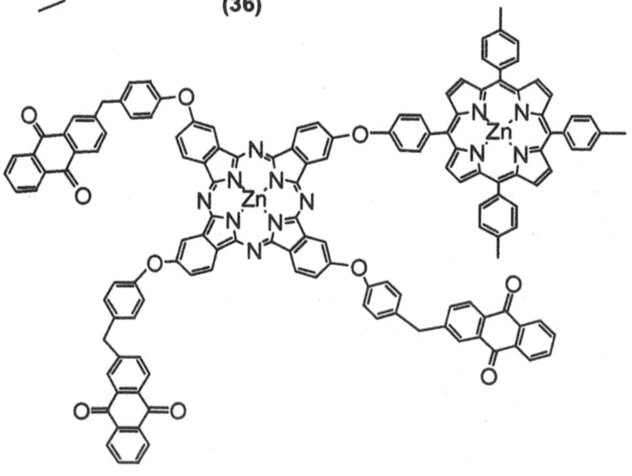

(37)

Fig. 3.64 The molecular structures of acceptor-linked phthalocyanines.

Some attempts at photoinduced hydrogen evolution with viologen-linked porphyrins are described in the following section.

3.5 Photoinduced Hydrogen Evolution with Viologen-linked Porphyrins

Photoinduced hydrogen evolution systems consisting of an electron donor (D), a photosensitizer (S), an electron carrier (C), and a catalyst have been used extensively for conversion of solar energy into chemical energy. In this reaction, charge separation between a photoexcited sensitizer and an electron carrier is one of the important steps. To improve this system some viologen-linked zinc porphyrins (S-C) have been synthesized. In the viologen-linked porphyrins, the photoexcited singlet state and the triplet state of porphyrin are more easily quenched by the bonded viologen compared with the viologen-free porphyrin. Since viologen-linked porphyrins can act as both a photosensitizer and an electron carrier in the same molecule, these compounds were applied to photoinduced hydrogen evolution, as shown in Scheme 3-18. In this section, studies of photoinduced hydrogen evolution with viologen-linked porphyrins are described.

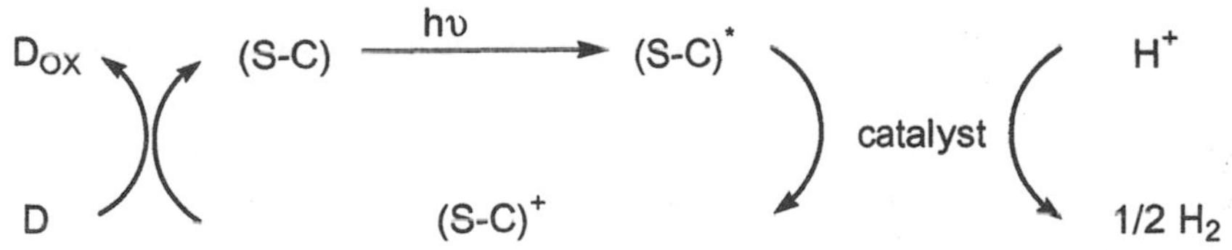

Scheme 3-18 Photoinduced hydrogen evolution using electron carrier-linked photosensitizer.
D: electron donor; S-C: electron carrier-linked photosensitizer.

R: p-tolyl

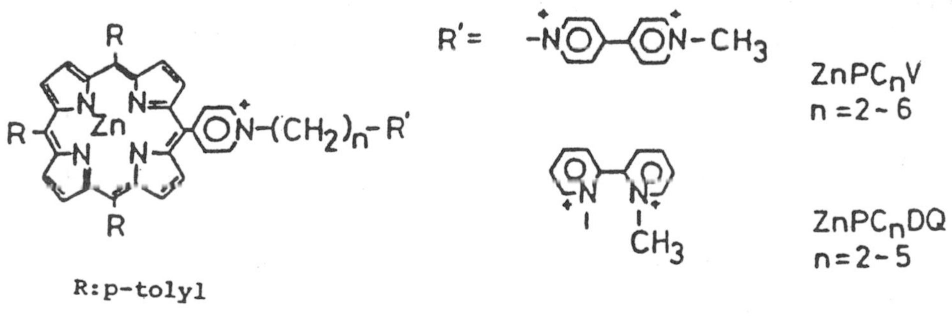

Fig. 3.65 The molecular structures of bipyridinium-linked porphyrins. (Reproduced with permission from *J. Chem. Soc. Chem. Commun.,* (**1986**) 170, The Royal Society of Chemistry)

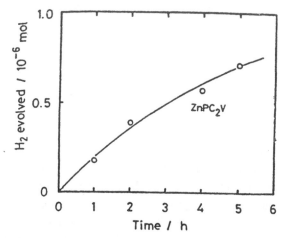

Fig. 3.66 Time dependence of the amount of hydrogen evolved by irradiation of the system containing NADPH (3.0 x 10⁻⁶mol), ZnPC₂V (9.9 x 10⁻⁸mol) and hydrogenase (1.0 ml) in 1.7 vol.% Triton X-100 solution. (Reproduced with permission from *J. Chem. Soc. Chem. Commun.,* (**1986**)170, The Royal Society of Chemistry)

Photoinduced hydrogen evolution has been attempted using viologen-linked zinc porphyrins (ZnPCₙV: n=2-6, Fig. 3.65).[71] Photoinduced hydrogen evolution is observed in the system containing NADPH, ZnPC₂V and hydrogenase in Triton X-100 solution, as shown in Fig. 3.66. However, no hydrogen evolution was observed using the other ZnPCₙV. The lifetime of the photoexcited triplet state is shorter in the case of ZnPC₂V than it is for the other ZnPCₙV. Transient absorption of reduced viologen or oxidized porphyrin is not observed in any case, indicating that intramolecular oxidative quenching of the excited triplet state of porphyrin to form the separated ion pair does not take place. The change in the spectrum of ZnPC₆V in the presence of NADPH by steady state irradiation is shown in Fig. 3.67. The absorption spectrum of ZnPC₆V changes with isosbestic points. This spectrum change is reversible, i.e., the new spectrum formed by irradiation disappeared by exposing the sample

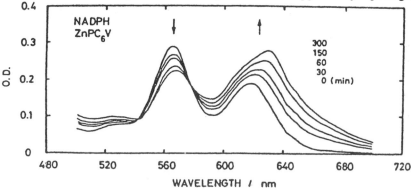

Fig. 3.67 Change in the spectrum of ZnPC₆V during steady state irradiation with NADPH (5.0 x 10⁻⁴mol dm⁻³) in 1.7 vol.% Triton X-100 solution. (Reproduced with permission from *J. Chem. Soc. Chem. Commun.,* (**1986**) 170, The Royal Society of Chemistry)

solution to the air, and the recovered spectrum was the same as that without irradiation. The above result shows that the absorption spectrum formed by irradiation corresponds to that of reduced porphyrin of $ZnPC_6V$. Therefore, irradiation of the solution containing NADPH and $ZnPC_6V$ causes reduction of the porphyrin of $ZnPC_6V$. The absorption spectrum of reduced viologen is not observed. Similar results have been obtained in the cases of $ZnPC_nV$ with n=3-5, showing that intramolecular electron transfer from reduced porphyrin to viologen does not take place, although the reduction of porphyrin proceeds under steady state irradiation in the presence of NADPH. Although the formation of reduced porphyrin is not observed in the case of $ZnPC_2V$, $ZnPC_2V$ is effective as both photosensitizer and electron carrier, as described above, indicating that the viologen of $ZnPC_2V$ is reduced under steady state irradiation with NADPH. In this case the reduction of the viologen does not take place by oxidative quenching of the photoexcited triplet state of porphyrin. The lifetime of the photoexcited triplet state of $ZnPC_2V$ is almost the same as that in the absence of NADPH, indicating that reductive quenching of the photoexcited triplet state of $ZnPC_2V$ does not proceed. From the fluorescence data, the intensity of $ZnPC_2V$ is lower than that of $ZnPC_nV$ with n=3-6. This suggests that quenching of the photoexcited singlet state of $ZnPC_2V$ takes place. $ZnPC_nV$ with n=3-6 is not effective for photoinduced hydrogen evolution. In these cases, intramolecular electron transfer from reduced porphyrin to viologen does not take place, even though porphyrin is reduced under steady state irradiation with NADPH. $ZnPC_nDQ$ with n=2-5 (2,2'-bipyridinium salt-linked zinc porphyrin) as shown in Fig. 3.68 are also used for photoinduced hydrogen evolution. In the case of $ZnPC_3DQ$, a trace amount of hydrogen is evolved, and in the case of $ZnPC_5DQ$ hydrogen evolution is observed by steady irradiation. $ZnPC_nDQ$ with n= 2 and 4 are not effective for photoinduced hydrogen evolution. Under steady state irradiation of $ZnPC_nDQ$ with NADPH, a spectrum change similar to that for the case of $ZnPC_nV$ is observed. In the case of $ZnPC_2DQ$, however, an irreversible spectrum change is observed and the compound is not

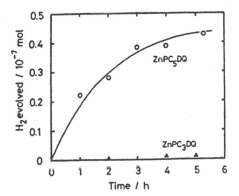

Fig. 3.68 Time dependence of the amount of hydrogen evolved by irradiation of the system containing NADPH (3.0×10^{-6} mol), $ZnPC_3DQ$ (2.6×10^{-7} mol) or $ZnPC_5DQ$ (2.7×10^{-7} mol) and hydrogenase (1.0 ml) in 1.7 vol. % Triton X-100 solution. (Reproduced with permission from *J. Chem. Soc. Chem. Commun.,* (**1986**) 170, The Royal Society of Chemistry)

Table 3.26 Photoinduced hydrogen evolution with NADPH-ZnP(CO$_2$)$_3$C$_n$V-hydrogenase system

Compound	Reaction time / h	Hydrogen evolved / 10^{-8} mol
ZnP(CO$_2$)$_3$C$_2$V	4	0.8
ZnP(CO$_2$)$_3$C$_3$V	2	1.0
ZnP(CO$_2$)$_3$C$_3$V	4	1.9
ZnP(CO$_2$)$_3$C$_4$V	4	1.6
ZnP(CO$_2$)$_3$C$_5$V	4	0.8
ZnP(CO$_2$)$_3$C$_6$V	4	1.4

(Reproduced with permission from *Bull. Chem. Soc. Jpn.*, **63** (1990) 2922)

effective for photoinduced hydrogen evolution, because of the instability of ZnPC$_2$DQ against irradiation.
Photoinduced hydrogen evolution with water-soluble viologen-linked zinc porphyrins (ZnP(CO$_2$)$_3$C$_n$V: n=2-6) and hydrogenase has been reported.[72] Photoinduced hydrogen evolution is observed in the system containing NADPH, ZnP(CO$_2$)$_3$C$_n$V, and hydrogenase. The amount of hydrogen evolved is shown in Table 3.26. It is apparent that all ZnP(CO$_2$)$_3$C$_n$V can be substrates of hydrogenase, and that every compound thus participates as both a photosensitizer and an electron carrier in the same molecule. The hydrogen evolution rate strongly depends on the methylene chain length; a lower hydrogen evolution rate is observed in the case of ZnP(CO$_2$)$_3$C$_2$V. An intramolecular electron transfer from porphyrin to viologen along the methylene chain may barely take place. When the methylene chain length is sufficiently long (n>3), the porphyrin ring possibly comes close enough to viologen so that an electron easily transfers directly from the porphyrin ring to viologen. A trace amount of hydrogen is observed when an individual component system containing ZnP(CO$_2$)$_3$CH$_3$, methyl viologen, NADPH, and hydrogenase is used, confirming that the linked viologen of ZnP(CO$_2$)$_3$C$_n$V serves as an electron carrier. However, the amount of hydrogen evolved is slight because of rapid back electron transfer between the reduced viologen to zinc porphyrin. Since the conformation of ZnP(CO$_2$)$_3$C$_n$V may be influenced by the environment, photoinduced hydrogen evolution is attempted in a surfactant micellar system. When a surfactant (Triton X-100) is added to a system containing ZnP(CO$_2$)$_3$C$_n$V, NADPH, and hydrogenase, and then irradiated, a remarkable increase in the hydrogen evolution is observed, as shown in Fig. 3.69. In the presence of Triton X-100, the fluorescence decay profile comprised two components with first-order decay: shorter and longer lifetimes are shown in Table 3.27. Although there are no remarkable differences in lifetime in the presence and in the absence of Triton X-100, the proportion of the shorter lifetime component increased in the presence of Triton X-100. The photoexcited singlet state with the shorter lifetime may play an important role in photoinduced hydrogen evolution,

Table 3.27 Fluorescence lifetime of ZnP(CO$_2$)C$_3$V

Solvent	τ_s / ns (A$_1$ / %)	τ_1 / ns (A$_2$ / %)
Triton X-100 (5vol %)	1.41 (78.3)	5.75 (21.7)
water	1.50 (68.8)	7.47 (31.2)

(Reproduced with permission from *Bull. Chem. Soc. Jpn.*, **63** (1990) 2922)

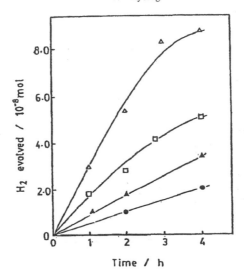

Fig. 3.69 Time dependence of hydrogen evolution. Sample solution (6.0 cm³) containing Zn-PC$_3$(C$_n$V) (2.0 x 10^{-6} mol dm^{-3}) NADPH (1.0 x 10^{-3} mol dm^{-3}) and hydrogenase (0.5cm³) was irradiated at 30°C in 5.0 vol % Triton X-100 / H$_2$O (△, □, ▲)and in H$_2$O (●). △ and ● : Zn-PC$_3$(C$_3$V), □ : Zn-PCPC$_3$(C$_4$V), ▲ : Zn-PCPC$_3$(C$_5$V). (Reproduced with permission from *Bull. Chem. Soc. Jpn.*, **63** (1990) 2922)

and rate increase in the presence of Triton X-100 may be caused by an increase in the component with the shorter lifetime.

Photoinduced hydrogen evolution with water-soluble viologen-linked cationic zinc porphyrins (ZnP(C$_n$V)$_4$) and hydrogenase is carried out under steady state irradiation. [81] An effective photoinduced hydrogen evolution system is achieved using ZnP(C$_4$V)$_4$ or ZnP(C$_5$V)$_4$, compared with an individual component system, as shown in Fig. 3.70. At the first stage of the photoreduction, the photoexcited triplet state of the porphyrin is reductively quenched by the electron donor, as the porphyrin used in this experiment is cationic. When ZnP(C$_n$V)$_4$ was irradiated in the presence of NADPH, the absorption spectrum intensity of 438 nm due to the Soret band of the porphyrin decreased, indicating that the reduced form of the porphyrin may be formed by the reductive quenching of the photoexcited triplet state of the porphyrin by NADPH. On the other hand, the bonded viologen is not reduced by NADPH. As shown in Fig. 3.71, the absorbance of the photoexcited triplet state of ZnP(C$_n$V)$_4$ strongly depends on the methylene chain length and the absorbance has the correlation of the photoinduced hydrogen evolution activity. The photoexcited triplet state of ZnP(C$_n$V)$_4$ corresponded to the initial hydrogen evolution rate. This result indicates that hydrogen evolution proceeds via the photoexcited triplet state of ZnP(C$_n$V)$_4$. The proposed mechanism is shown in Scheme 3-19.

In the first step, the photoexcited singlet state of ZnP(C$_n$V)$_4$ is formed by the irradiation. Then the photoexcited triplet state of ZnP(C$_n$V)$_4$ is formed by an intercrossing reaction in the second step. In the third, the porphyrin moiety is reduced by quenching of the photoexcited triplet state with NADPH. Finally the electron transfer from the reduced porphyrin moiety to the viologen occurs and hydrogen evolves by electron transfer from reduced viologen to hydrogenase. According to fluorescence lifetime

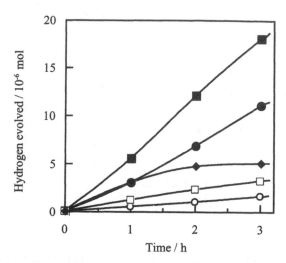

Fig. 3.70 Time dependence of hydrogen evolution. A sample solution containing (4.0 ml) ZnP(C V) $(2.5 \times 10^{-6}$ mol dm^{-3}), NADPH $(2.0 \times 10^{-3}$ mol dm^{-3}) and hydrogenase (0.35 unit) is irradiated at 30 °C. F: ZnTMPyP $(2.5 \times 10^{-6}$ mol dm^{-3}), NADPH $(2.0 \times 10^{-3}$ mol dm^{-3}), methyl viologen $(1.0 \times 10^{-5}$ mol dm^{-3}) and hydrogenase (0.35 unit) is irradiated at 30°C. (O) n=3; (●) n=4; (■) n=5; (□) n=6. (Reproduced with permission from *J. Porphyrins Phthalocyanines,* **2** (1998) 201, John Wiley & Sons Limited)

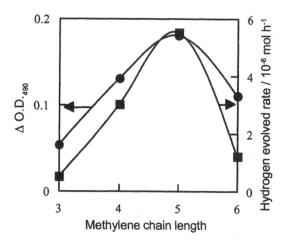

Fig. 3.71 The relationship of absorption of the photoexcited triplet state of ZnP(C$_n$V)$_4$ and the initial rate of hydrogen evolution. The absorption of the photoexcited triplet state of ZnP(C$_n$V)$_4$ was measured at 490 nm. ●:absorbance of the photoexcited triplet state of ZnP(C$_n$V)$_4$; ■:initial rate of hydrogen evolution with hydrogenase and ZnP(C$_n$V)$_4$. (Reproduced with permission from *J. Porphyrins Phthalocyanines,* **2** (1998) 201, John Wiley & Sons Limited)

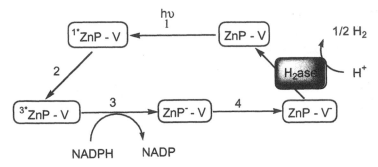

Scheme 3-19 Proposed mechamism of hydrogen evolution. (Reproduced with permission from *J. Porphyrins Phthalocyanines*, **2** (1998) 201)

decay measurements, the electron transfer from the photoexcited singlet state of porphyrin to the bonded viologen is very rapid. This result indicates that no photoinduced hydrogen evolution occurs via the photoexcited singlet state of porphyrin. The photoexcited triplet state of $ZnP(C_nV)_4$ corresponds to the initial hydrogen evolution rate, indicating that hydrogen evolution proceeds via the photoexcited triplet state of $ZnP(C_nV)_4$.

Photoinduced hydrogen evolution with water-soluble bisviologen-linked cationic zinc porphyrin ($ZnP(C_4V_AC_4V_B)_4$) and hydrogenase is also carried out under steady state irradiation.[83] Using $ZnP(C_4V_AC_4V_B)_4$, the effective photoinduced hydrogen evolution system is accomplished as shown in Fig. 3.72.

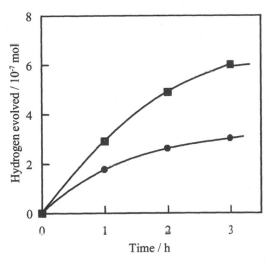

Fig. 3.72 Time course of photoinduced hydrogen evolved using $ZnP(C_4V_AC_4V_B)_4$ (■) under steady state irradiation. [NADPH]=2.0 mmol dm^{-3}, [viologen-linked zinc porphyrin]=2.5 mmol dm^{-3} and H$_2$ase=0.35 unit.●: $ZnP(C_4)_4$ and butylviologen ([NADPH]=2.0 mmol dm^{-3}, [$ZnP(C_4)_4$]=2.5 mmol dm^{-3}, [MV^{2+}]=10 mmol dm^{-3} and H$_2$ase=0.35 unit). (Reproduced with permission from *J. Mol. Catal. A: Chem.*, **120** (1997) 5, Elsevier Science)

Although photoinduced hydrogen evolution using viologen-linked porphyrins is observed, effective hydrogen evolution systems have not yet been achieved. An effective system using viologen-linked porphyrin will be achieved by making the following improvements: 1) high quantum yield for the formation of charge-separated species between the reduced viologen and porphyrin moiety and 2) suppression of back electron transfer between the reduced viologen and porphyrin moiety.

The synthesis of compounds with these requirements is necessary to establish a highly efficient photoinduced hydrogen evolution system using viologen-linked porphyrins.

3.5.1 Application of phthalocyanines

As phthalocyanines absorb light of longer wavelength (600-700 nm), these compounds are also suitable photosensitizers for photochemical reactions. In this section, the photoreduction of viologens using metallophthalocyanine is described. The photoreduction system using metallophthalocyanine is also expressed in Scheme 3-20. Table 3.28 shows the quantum efficiency of photoreduction of methyl viologen using various metallophthalocyanines in a DMF/water solution. In the system using

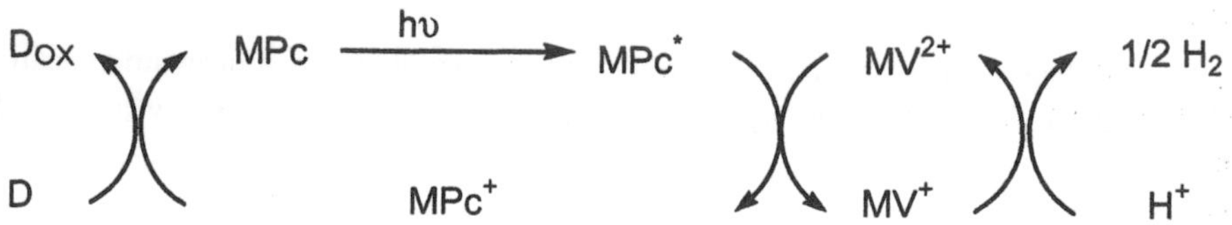

Scheme 3-20 Photoinduced hydrogen evolution using metallophthalocyanine (MPc). D: Electron donor.

Table 3.28   Quantum yield of methyl viologen photoreduction

Phthalocyanine	Φ_{MV^+}
MgPc	0.27
MgPc(OCH$_3$)$_4$	0.34
MgPc(C$_6$H$_5$)$_4$	0.23
MgPc(NO$_2$)$_4$	0.22
AlClPc	0.98
ZnPc	7.2×10^{-2}
Na$_2$Pc	5.1×10^{-2}
NiPc	3.0×10^{-2}
CaPc	2.5×10^{-2}
Li$_2$Pc	8.7×10^{-3}
CuPc	2.2×10^{-3}
FePc	9.0×10^{-4}

metallophthalocyanine, the composition of the solvent is an important factor. In the case of MgPc, no accumulation of the reduced methyl viologen is observed in the DMSO/water (=1 / 1) solution. In the case of DMSO/water (=95 / 5) solution, quantum efficiency of photoreduction of methyl viologen is estimated to be 0.47. The photoreduction of methyl viologen using a system containing MgPc and an electron donor occurs via oxidative quenching.[86] Among metallophthalocyanines, AlPcCl has the higher activity for photoreduction of methyl viologen and the quantum efficiency of the photoreduction of methyl viologen is estimated to be 0.98. In the case of a system containing AlPcCl and an electron donor, the photoreduction of methyl viologen occurs via reductive quenching. In the cases of system containing metallophthalocyanines except AlPcCl, on the other hand, the photoreduction of methyl viologen occurs via oxidative quenching.[87]

The photoexcited singlet state and triplet state of $MgPc(OCH_3)_4$ is able to reduce methyl viologen.[87] The quenching rate constant of fluorescence of MgPc by methyl viologen is estimated to be 4.7×10^{10} $mol^{-1}dm^3 s^{-1}$. On the other hand, the quenching rate constant of the photoexcited triplet state of MgPc by methyl viologen is estimated to be 6.7×10^7 $mol^{-1}dm^3 s^{-1}$.

In the system using metallophthalocyanine, little photoinduced hydrogen evolution is observed using platinum colloid or hydrogenase. Photoinduced hydrogen evolution was observed using a NiPc system with sputtering platinum-methyl viologen for the first time.[87]

As water-soluble zinc phthalocyanines form dimers in aqueous solution, these compounds are not suitable for the photoreduction of methyl viologen. By the addition of a surfactant such as hexadecyltrimethylammonium bromide (CTAB) or Triton X-100, however, the formation of dimers is suppressed. For example, a photoreduction of methyl viologen is accomplished using a system containing zinc tetrasulfonatophthalocyanine ($ZnPcSO_4$) and cysteine (electron donor) in surfactant micellar solution.[88]

Many attempts have been made using additional doping agents to improve the electrical performances of molecular solar cells. MgPc powders have been coated with a film of air-oxidized tetramethyl-*p*-phenylenediamine by evaporating an acetone solution of the amine. [91-93] In the absence of a dopant, no photovoltage under irradiation is observed. Surface doping induces a photovoltage of 200 mV under strong illumination conditions (500 W lamp). Similar effects have been described for other devices.[94-99] Sandwiches such as SnO_2 / MPc / organic dyes / Indium have been realized (Fig. 3.73). The energy conversion efficiencies of these devices are remarkably good and reach 0.5% for white light under natural condition of irradiation (70 mW cm^{-2}).[97]

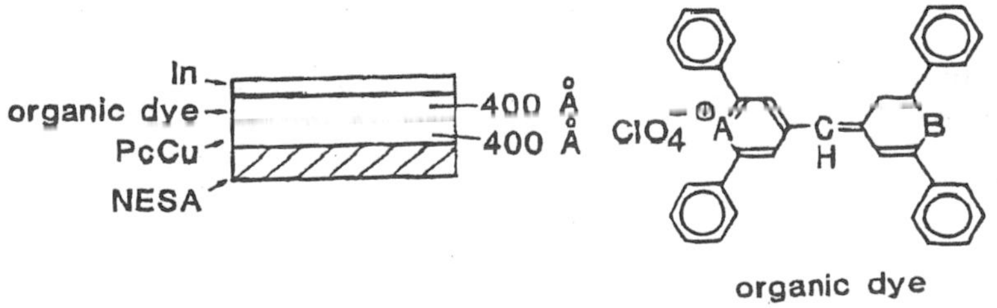

Fig. 3.73 Organic solar cells made of two different dyes.(Reproduced with permission by F.J. Kampas, K. Yamashita, J. Fajer, *Nature,* **284** (1980) 40, Macmillan Magazines Limited)

A comparison of the main systems exhibiting the higher conversion efficiencies described in the literature is shown in Table 3.29.[102] The influence of the complexed metal ions on the performances of the organic solar cells strictly follows their O_2-binding ability. The order of decreasing efficiency, it is as follows[103]: MgPc > BePc > FePc > CuPc> CoPc. Metalloporphyrin complexes show a very similar correlation:[104]: MgTPP > ZnTPP> CuTPP> H_2TPP > CoTPP.

The power conversion efficiency of molecular solar cells is still limited by the electrical field dependence of the quantum efficiency of charge-carrier generation. This limitation is clearly correlated with the very low overall conductivity of the molecular materials and the space-charge limitations which correspondingly arise.

Table 3.29　Power conversion efficiencies of the best organic solar cells based on metallophthalocyanine semiconductors published in the literature

System	Efficiencies originally described (%) and corresponding exp. conditions	Standardized Yields %
Ag/PcMg/Al	1.5×10^{-3}; no FF white light; 0.5 mW/cm^2	1.2×10^{-4}
Ag/PcMg/Al	10^{-3}; white light;	?
NESA/x-PcH$_2$/Al	0.02; monochromatic light (670 nm); 100 mW/cm^2; corrected for transmission	1.7×10^{-3}
Au/PcH$_2$/Al	2.5×10^{-4}; monochromatic light (6328 Å); 67 mW/cm^2	2.1×10^{-5}
Au/PcZn/In	0.023; same conditions	1.9×10^{-3}
?/(PcAl)$_2$Se/?	3; unfiltered light;	?
?/(PcV)$_2$O/?	1-100 lux	
Au/PcZn/Cr		7.6×10^{-3}
Au/PcZn/In		5.9×10^{-3}
Au/PcMg/Cr		1.2×10^{-2}
Au/PcMg/In-Al		1.9×10^{-2}
?/PcInCl/?	0.5; 10 mW/cm^2	?
NESA/x-PcH$_2$/In	1.2%; solar simulated; corrected for transmission	? 1.2×10^{-2}

(Reproduced with permission by F.J. Kampas, K. Yamashita, J. Fajer, *Nature,* **284** (1980) 40, Macmillan Magazines Limited)

References

1. J.R. Darwent, P. Douglas, A. Harriman, G. Porter and M.-C. Richoux, *Coord. Chem. Rev.,* **44** (1982) 83.
2. C. Laane, I. Willner, J.W. Otvos and M. Calvin, *Proc. Natl. Acad. Sci.* USA, **78** (1981) 5928.
3. K. Kalyanasundaram, J. Kiwi and G. Gratzel, *Helv. Chim. Acta,* **61** (1978) 2720.
4. M. Kaneko, N. Takahashi, Y. Yamauchi and A. Yamada, *Bull. Chem. Soc. Jpn.,* **57** (1984) 156.
5. A. Fujishima and K. Honda, *Nature,* **238** (1972) 37.
6. P. Salvador, C. Gutierrez, G. Campet and P. Hagenmiiller, *J. Electrochem. Soc.,* **131** (1984) 550.
7. K. Kalyanasundaram, B. Borgarello and G. Gratzel, *Helv. Chim. Acta,* **64** (1981) 362.
8. K.K. Rao, I.N. Gogotov and D.O. Hall, *Biochimie,* **60** (1978) 291.
9. K.K. Rao, L. Rosa and D.O. Hall, *Biochem. Biophys. Res. Commun.,* **68** (1976) 21.
10. I. Fry and G. Papageorgiou, *Z. Naturforsch., Teil C,* **32** (1977) 110.
11. D.A. Lappi, F.E. Stoizenbach, N.A. Kaplan and M.D. Kamen, *Biochem. Biophys. Res. Commun.,* **69** (1976) 878.
12. E. Greenbaum, R.R.L. Gullard and W.G. Gunda, *Photochem. Photobiol.,* **37** (1983) 649.
13. O. Johansen, A. Launikonis, A.W.-H. Mau and W.H.F. Sasse, *Aut. J. Chem.,* **33** (1980) 1643.
14. P. Keller and A. Moradpour, *J. Am. Chem. Soc.,* **102** (1980) 7193.
15. A.A. Krasnovski, G.P. Brin, I.P. Gotov and V.P. Pschepkov, *Dokl. Acad. Nauk. SSSR.* **225** (1975) 711.
16. A.A. Krasnovski, G.P. Brin, and U.V. Nikandrov, *Dokl. Acad. Nauk. SSSR.* **228** (1975) 1214.
17. J.A.M. Smith and D. Mauzerall, *Photochem. Photobiol.,* **34** (1981) 407.
18. I. Okura and N. Kim-Thuuan, *J. Mol. Catal.,* **6** (1979) 227.
19. G. McLendon and D.S. Miller, *J. Chem. Soc., Chem. Commun.* (**1980**) 553.
20. A. Harriman, G. Porter and M.-C. Richoux, *J. Chem. Soc., Faraday Trans.2,* **77** (1981) 833.
21. N. Kim-Thuuan, Doctoral Thesis, 1980, Tokyo Institute of Technology.
22. I. Okura, M. Takeuchi and N. Kim-Thuuan, *Photochem. Photobiol.,* **33** (1981) 413.
23. R.H. Schmehl and D.G. Whitten, *J. Phys. Chem.,* **85** (1981) 3473.
24. M. Rougee, T. Ebbesen, F. Ghetti and R.V. Besasson, *J. Phys. Chem.,* **86** (1982) 4404.
25. I. Okura, N. Kaji, S. Aono, T. Kita and A. Yamada, *Inorg. Chem.,* **24** (1985) 451.
26. A. Harriman, G. Porter and M.-C. Richoux, *J. Chem. Soc., Faraday Trans.2,* **77** (1981) 833.
27. E. Amouyal and B. Zidler, *Isr. J. Chem.,* **22** (1982) 117.
28. I. Willner, Jer-Ming Yang, C. Laane, J.W.Otvos and M. Calvin, *J. Phys. Chem.,* **85** (1981)

3277.

29. P. Neta, *J. Phys. Chem.,* **85** (1981) 3678.

30. V. Balzani and F. Scandola, *in Energy Resources through Photochemistry and Catalysis* (1983) ed. M. Gratzel, Academic Press, New York.

31. I. Okura, M. Takeuchi, S. Kusunoki and S. Aono, *Chem. Lett.,* (**1982**) 187.

32. M. Shin, *Methods in Enzymology* **23** (1971) 440.

33. C.K. Gratzel and M. Gratzel, *J. Phys. Chem.,* **86** (1982) 353.

34. P.A. Brugger, P.P Infelta, A.M. Braun and M. Gratzel, *J. Am. Chem. Soc.,* **103** (1981) 320.

35. L.A. Kelly and M.A.J. Rodgers, *J. Phys. Chem.,* **98** (1994) 6386.

36. P.J.G. Coutinho and S.M.B. Costa, *J. Photochem. Photobiol. A. Chem.,* **82** (1994) 149.

37. Y. Amao and I. Okura, *J.Mol. Catal. A: Chem.,* **103** (1995) 69.

38. Y. Amao and I. Okura, *J.Mol. Catal. A: Chem.,* **105** (1996) 125.

39. Y. Amao and I. Okura, *Porphyrins,* **4** (1995) 129.

40. J. Deisserhofer, O.Epp, K. Miki, R. Huber, H. Michel, *Nature,* **318** (1985) 618.

41. Y. Sakata, Y. Nakashima, Y. Goto, H. Tatemitsu, S. Misumi, T. Asahi and N. Mataga, *J. Am. Chem. Soc.,* **111** (1989) 8970.

42. Y. Sakata, H. Tsue, Y. Goto, S. Misumi, T. Asahi, S. Nishikawa, T. Okada and N. Mataga, *Chem. Lett..,* (**1991**) 1307.

43. D. Gust, T.A. Moore, P.A. Liddell, G.A. Nemeth, L.R. Makings, A.L. Moore, D. Barrett, P.J. Pessiki, R.V. Bensasson, M. Rougee, C. Chachaty, F.C. De Scchryver, M. Van der Auweraer, A.R. Holzwarth and J. S. Connolly, *J. Am. Chem. Soc.,* **109** (1987) 846.

44. T.A. Moore, D. Gust, P. Mathis, J.-C. Mialocq, C. Chachaty, R.V. Bensasson, E.J. Land, D. Doizi, P.A. Liddell, W.R. Lehman, G.A. Nemeth and A.L. Moore, *Nature,* **307** (1984) 630.

45. D. Gust and T.A. Moore, *Science,* **244** (1989) 35.

46. D. Gust, T.A. Moore, D. Barrett, L.O. Harding, L.R. Makings, P.A.Liddell, F.C. De Scchryver, M. Van der Auweraer, R.V. Bensasson and M. Rougee, *J. Am. Chem. Soc.,* **110** (1988) 321.

47. D. Gust, T.A. Moore, A.L. Moore, L.R. Makings, G. Seely, X. Ma, T.T. Trier and F. Gao, *J. Am. Chem. Soc.,* **110** (1988) 7567.

48. D. Gust, T.A. Moore, A.L. Moore, S.-J. Lee, E. Bittersmann, D.K. Luttrull, A.A. Rehms, J.M. DeGraaziano, X.C.Ma, F. Gao, R.E. Belford and T.T. Trier, *Science,* **248** (1990) 199.

49. M. Ohkohchi, A. Takahashi, N. Mataga, T. Okada, A. Osuka, H. Yamada and K. Maruyama, *J. Am. Chem. Soc.,* **115** (1993) 12137.

50. A. Osuka, S. Nakajima, K. Maruyama, N. Mataga, T. Asahi, I. Yamazaki, Y. Nishimura, T. Ohno and K. Nozaki, *J. Am. Chem. Soc.,* **115** (1993) 4577.

51. A. Osuka, S. Nakajima, K. Maruyama, N. Mataga and T. Asahi, *Chem. Lett.,* **(1991)** 1003.

52. A. Osuka, S. Maruo, K. Maruyama, N. Mataga, Y. Tanaka, S. Taniguchi, T. Okada, I. Yamazaki and Y. Nishimura, *Bull. Chem. Soc. Jpn.,* **68** (1995) 262.

53. J.W. Arbogast, A.P. Darmanyan, C.S. Foote, Y. Rubin, F.N. Diederich, M.M. Alvarez, S.J. Anz, R.L. Whetten, *J. Phys. Chem.,* **95** (1991) 11.

54. C. Reber, L.Yee, J. McKiernan, J.I. Zink, R.S. Williams, W.M. Tong, D.A.A. Ohlberg, R.L. Whetten, F. Diederich, *J. Phys. Chem.,* **95** (1991) 2127.

55. K. Palewska, J. Sworakowski, H. Chojnacki, E.C. Meister, U.P. Wild, *J. Phys. Chem.,* **97** (1993) 12167.

56. J.L. Anderson, Y.-Z. An, Y. Rubin, C.S. Foote, *J. Am. Chem. Soc.,* **116** (1994) 9763.

57. S.-K. Lin, L.-L.Shiu, K.-M. Chien, T.-Y. Luh, T.-I. Lin, *J. Phys. Chem.,* **99** (1995) 105.

58. R.R. Hung, J.J. Grabowski, *J. Phys. Chem.,* **95** (1991) 6073.

59. Y. Zeng, L. Biczok, H. Linschitz, *J. Phys. Chem.,* **96** (1992) 5237.

60. D. Gust, T.A. Moore, A.L. Moore, L. Leggett, S. Lin, J.M. DeGraziano, R.M. Hermant, D. Nicodem, P. Craig, G.R. Seely and R.A. Nieman, *J. Phys. Chem.,* **97** (1993) 7926.

61. H. Imahori, K. Hagiwara, T. Akiyama, S. Taniguchi, T. Okada, M. Shirakawa and Y. Sakata, *Chem. Lett.,* **(1995)** 265.

62. H. Imahori, K. Hagiwara, M. Aoki, T. Akiyama, S. Taniguchi, T. Okada, M. Shirakawa and Y. Sakata, *J. Am. Chem. Soc.,* **118** (1996) 11771.

63. T. Drovetskaya, C.A. Reed and P. Boyd, *Tetrahedron Lett.,* **36** (1995) 7971.

64. D. Kuciauskaya, S. Lin, G.R. Seely, A.L. Moore, T.A. Moore, D. Gust, T. Drovetskaya, C.A. Reed and P. Boyd, *J. Phys. Chem.,* **100** (1996) 15926.

65. G. Blondeel, D. Keukeleire, A. Harriman, L.R. Milgrom, *Chem. Phys. Lett.,* **118** (1985) 77.

66. R.K. Force, R.J. McMahon, J. Yu and S. Wrighton, *Spectrochimica Acta,* **45** (1989) 23.

67. S. Noda, H. Hosono, I. Okura, Y. Yamamoto and Y. Inoue, *J. Chem. Soc. Faraday Trans. 1,* **86** (1990) 811.

68. Y. Kinumi and I. Okura, *Inorg. Chim. Acta,* **152** (1988) 77.

69. J. Hirota, T. Takeno and I. Okura, *J. Photochem. Photobiol. A: Chem.,* **77** (1994) 29.

70. J. Hirota and I. Okura, *J. Phys. Chem.,* **97** (1993) 6867.

71. S. Aono, N. Kaji and I. Okura, *J. Chem. Soc. Chem. Commun.,* **(1986)** 170.

72. I. Okura and Y. Kinumi, *Bull. Chem. Soc. Jpn.,* **63** (1990) 2922.

73. I. Okura and H. Hosono, *J. Phys. Chem.,* **97** (1993) 6867.

74. I. Okura, N. Kaji, S. Aono, T. Kita and A. Yamada, *Inorg. Chem.,* **24** (1985) 451.

75. Y. Yamamoto, S. Noda, N. Nanai, I. Okura and Y. Inoue, *Bull. Chem. Soc. Jpn.,* **64** (1991) 1392.

76. J. D. Hopfield, *Proc. Nat. Acad. Sci. USA,* **71** (1974) 3640.

77. E.E. Batova, P.P Levin, and V.Y. Shafirovich, *New J. Chem.,* **14** (1990) 269.

78. Y. Amao, T. Hiraishi and I. Okura, *J. Mol. Catal. A:Chem.,***126** (1997) 13.

79. Y. Amao, T. Hiraishi and I. Okura, *J. Mol. Catal. A:Chem.,***126** (1997) 21.

80. Y. Amao, T. Kamachi and I. Okura, *J. Photochem. Photobiol. A: Chem.,* **98** (1996) 59.

81. Y. Amao, T. Kamachi and I. Okura, *J. Porphyrins Phthalocyanines,* **2** (1998) 201.

82. Y. Amao, T. Kamachi and I. Okura, *Inorg. Chim. Acta,* **267** (1998) 257.

83. Y. Amao, T. Kamachi and I. Okura, *J. Mol. Catal. A: Chem.* **120** (1997) 5.

84. J.X. Liu, Q. Yu, Q.F. Zhou and H.J. Xu, *J. Chem. Soc. Chem. Commun.* (**1990**) 260.

85. L. Li, S. Shen, Q. Yu, Q.F. Zhou and H.J. Xu, *J. Chem. Soc. Chem. Commun.* (**1991**) 619.

86. T. Tanno, D. Woehrle, M. Kaneko and A. Yamada, *Ber. Bunsenges. Phys. Chem.,* **84** (1980) 1032.

87. H. Ohtani, T. Kobayashi, T. Tanno, A. Yamada, D. Woehrle and T. Ohno, *Photochem. Photobiol.,* **44** (1986) 125.

88. J.R. Darwent, *J. Chem. Soc., Chem. Commun.,* (**1980**) 805.

89. M. Calvin and D.R. Kearns, *US Patent,* 3,057,947 CA 2990a (1963).

90. D.R. Kearns and M. Calvin, *J. Chem. Phys.,* **29** (1958) 950.

91. H. Meier, *Umshau. Wiss. Tech.,* **66** (1966) 438 (in German).

92. M.I. Fedorov and V.A. Benderskii, *Sov. Phys. Semicond.,* **4** (1971) 1198.

93. G.G. Komissarov, Yu-S. Shumov and Yu-E. Benderskii, *Dokl. Akad. Nauk., SSSR,* **187** (1969) 670.

94. M.E. Fedorov and V.A. Benderskii, *Fiz. Tekh. Poluprov,* **4** (1970) 1403 (in Russian).

95. M.I. Fedorov, E.P. Zinov'eva, V.A. Shorin and L.I. Nutrikhina, *Izv. Vyssh. Uch. Zav. Fiz.* **20** (1977) 159 (in Russian).

96. M.I. Fedorov, E.P. Zinov'eva, V.N. Shashaurov and L.I. Nutrikhina, *Izv. Vyssh. Uch. Zav. Fiz.* **20** (1977) 157 (in Russian).

97. G.A. Stepanova, L.S. Volkova, M.A. Gainullina and F.F. Yumakulova, *Zh. Fiz. Khim.* **51** (1977) 1771.

98. R.O. Loutfy and J.H. Sharp, *J. Chem. Phys.,* **71** (1979) 1211.

99. R.O. Loutfy, J.H. Sharp, C.K. Hsiao and R. Ho, *J. Appl. Phys.,* **52** (1981) 5218.

100. M. Martin, J.-J. Andre and J. Simon, *Nouv. J. Chim.,* **5** (1981) 485.

101. V.A. Ilatovskii, G.G. Komissarov, *Zh. Fiz. Khim.* **49** (1975) 1352 (in Russian).

102. F.J. Kampas, K. Yamashita and J. Fajer, *Nature,* **284** (1980) 40.

103. D.L. Morel, *Mol. Cryst. Liquid. Cryst.,* **50** (1979) 127.

104. D.L. Morel, A.K. Ghosh, T. Feng, E.L. Stogryn, P.E. Purwin, R.F. Shaw and C. Fishman, *Appl. Phys. Lett.,* **32** (1978) 495.

105. A.K. Ghosh and T. Feng, *J. Appl. Phys.,* **49** (1978) 5982.

106. J. Fabian and H. Hartmann, *Light Absorption of Organic Colorants,* Berlin, Springer Verlag, 1980.

Chapter 4   Optical Sensors

4.1   Fundamental Aspects of Optical Oxygen Sensors

Oxygen is an immensely important chemical species, essential for life. The determination of oxygen levels is required in many various fields. In environmental analysis, for example, oxygen measurement provides an indispensable guide to the overall condition of the ecology, and it is routine practice to monitor oxygen levels continuously in the atmosphere and in water. In the medical field, oxygen levels in exhaled air or in the blood of a patient are key physiological parameters for judging health. Such parameters should ideally be monitored continuously. The measurement of oxygen levels is also essential in industries which utilize metabolizing organisms: yeast for brewing and bread making, and the plants and microbes that are used in modern biotechnology, such as those producing antibiotics and anticancer drugs. In this section, the background of oxygen measurements is described and the work to develop new optical oxygen sensors which utilize the luminescence of metal complexes is discussed.

Several methods for oxygen detection have been reported so far, some based on titration,[1] amperometric,[2] chemiluminescence[3] or thermoluminescence.[4] Winkler titration has been employed for the measurement of oxygen for many years and is considered, to some extent, to be the standard method.[1] However, the time-consuming and cumbersome nature of the titration has hindered its application to process monitoring. Clark-type electrodes provided a breakthrough technique for the measurement of oxygen and have become the conventional method. The Clark cell is robust and quite reliable. However, it is bulky and not readily miniaturized without great expense. Since the cells are based on the reduction of oxygen at the cathode and the diffusion-limited passage of oxygen through the membrane, any factor that can change the diffusion resistance, such as fouling of the membrane or a change in flow conditions in the testing fluid, can cause measurement error.[5,6] They also suffer from electrical interference and since they consume oxygen, can easily generate misleading data. Thus, there is a constant interest in new, superior techniques for oxygen detection, and optical oxygen sensors represent one of the new hopefuls in this area. Optical oxygen sensors should be inexpensive, easily miniaturized and simple to use. They should not suffer from electrical interference or consume oxygen. Many of the chemical sensors developed to measure oxygen employ luminescence as the principle of the measurement. Optical sensors based on photoluminescence quenching usually measure the luminescence intensity[7,8] or lifetime[9] of organic dye immobilized in a polymer film that varies by the concentration of the analyte. However, in the development of the optical oxygen sensor, the detection method as well as the choice of dye and matrix material and the film preparation method are crucial factors.[10]

Recent developments in optical oxygen sensors have been reviewed.[11] In 1985, a new optical method based on phosphorescence lifetime quenching was introduced for the measurement of oxygen in biological systems.[12] This technique seemed to be very attractive because it did not suffer many of the practical limitations of the fluorescence intensity quenching methods used by previous workers.[13,14] Also, the relatively long phosphorescence lifetime of the materials employed allowed the use of less sophisticated measurement instrumentation. The possibility of using this technique to develop a new gaseous oxygen sensor based on membrane immobilization of organic dyes, especially the room-temperature phosphorescent metalloporphyrins, has been explored. Most optical oxygen sensors respond specifically and reversibly to molecular oxygen by a change in the intensity of luminescence emitted from a probe molecule.

As shown in the Jablonski diagram (Fig. 2.3) luminescence (spontaneous emission of light) is produced in certain molecules following excitation by light from their ground state to higher energy states. When the excited molecule returns to the ground state, it emits light (fluorescence or phosphorescence) at a longer wavelength than the original simulated light (Stokes shift). Fluorescence is a short-lived emission from the singlet state with the electrons spin-paired. As the transition from the singlet state to the ground singlet state is spin-allowed, i.e. with no change in multiplicity it has a high probability of occurrence and the decay time of fluorescence is usually short, 10^{-9} to 10^{-7} s, or the same order as the lifetime of an excited singlet state. On the other hand, phosphorescence involves a change in electron spin, so the transition from the lowest excited triplet state to the singlet ground state occurs at a low probability and phosphorescence is a long-lived emission (about 10^{-5} to 10 s).[15,16] Also, owing to the relative energies of the excited states, phosphorescence occurs at longer wavelengths than fluorescence. When the molecules in the triplet state have long lifetimes, they are particularly susceptible to interactions with other molecules. The quenching by oxygen is diffusion-limited and can be described by the Stern-Volmer relationship.[17] This is valid for both phosphorescence and fluorescence lifetimes and intensities.

$$I_0 / I = \tau_0 / \tau = 1 + k_q \tau_0 [O_2] \tag{4-1}$$

where, I_0 and I are the luminescence intensities, and τ_0 and τ are lifetimes in the absence of oxygen and in the presence of oxygen (concentration $[O_2]$), respectively. The quenching rate constant, k_q, is dependent on the physical constants of the system and is described by a modified Smoluchowski equation[18-20]:

$$k_q = 4pNp(D_A + D_B) \times 10^3 \tag{4-2}$$

where N is the Avogadro number, p is a factor related to the probability of each collision causing quenching and to the radius of interaction between the donor and the oxygen. D_A and D_B are the diffusion coefficients for the donor and acceptor, respectively. In general, under defined experimental conditions, k_q can be considered constant. Using eq. (4-1), it is possible to make intensity or lifetime measurements of phosphorescence (or fluorescence) then relate this measurement to the oxygen concentration.

Kautsky has discovered that the phosphorescence[21] and the fluorescence[22] of surface-adsorbed dyes such as trypaflavin, benzoflavin, safranin, chlorophyll and porphyrins are sensitively quenched by molecular oxygen, which, in its electronic ground state, is a triplet molecule. This led to the design of the first optical sensor for oxygen. Depending on the measurement of either phosphorescence or fluorescence, extremely low sensitivities (typical detection limit in phosphorimetry 0.067 Pa oxygen) are achieved.[23] Zhakharov and Grishaeva[24] have utilized this effect to devise an optical-sensor for low oxygen levels as they occur in wastewater. Detection limits are reported to be as low as 0.5 mg oxygen per liter. The fast growing interest in optical sensors has led to a variety of other sensor types in the past. Bergman[25] and others[26-30] have used polycyclic aromatic hydrocarbons (PAHs) which are found to be efficiently quenched by oxygen in the 0-40 kPa range. The PAHs are either dissolved in a polymer,[27-30] soaked into porous glass,[25] or covalently immobilized on glass support.[26] Peterson et al., by combining Kautsky's adsorption technique with the sensitivity of the PAHs, developed a fiberoptic sensor for oxygen based on the oxygen-sensitive fluorescence of Amberlite-adsorbed perylene dibutyrate.[13]

Changes in luminescence intensity can therefore be used to monitor oxygen.[14] It is also shown that the fluorescent dye pyrene butyric acid (PBA) might be used to determine oxygen in biological microenvironments. Subsequently, PBA is used by several workers for oxygen measurement in tissue.[31,32] PBA is also used in a gaseous oxygen sensor,[28] with a thin layer of the dye (10 μm thick) trapped behind a 6-μm thick gas-permeable Teflon membrane. For *in vivo* oxygen concentration measurement, a catheter tip sensor that can be used in tissue or introduced into blood vessels has been developed.[5] The fluorescence dye (perylene dibutyrate) used in this system is held behind a 25-μm thick porous polypropylene gas-permeable membrane. The oxygen concentration is calculated using an equation derived from the Stern-Volmer relationship, including a correction for the non-linearity of the response.

Recently, phosphorescence quenching-based sensors have increasingly become the focus of attention as they have several advantages over fluorescence-based ones. The longer excited state lifetimes of phosphorescent indicators give rise to high quenching efficiently by oxygen. In addition, the long excitation and emission wavelengths are more compatible with the available optical monitoring technology. Historically, the first optical sensors for oxygen described in the literature are based on room-temperature phosphorescence quenching of immobilized dyes, though these dyes tended to be photolabile.[33] Indeed, this tends to be a general problem associated with excited triplet state molecules and is one of the major drawbacks associated with their use. As molecules in the triplet state have long lifetimes, they should in principle form the basis of an extremely sensitive oxygen sensor. However, until recently, there have been few molecules that exhibit a strong phosphorescence yield at room temperature. Recently, several probes that are suitable for room temperature phosphorescence applications have been presented. More recent room-temperature phosphorescence quenching-based sensors have utilized metal chelates such as tetrakis(pyrophosphito)diplatinate (II) [34] and 8-hydroxy-7-iodo-5-quinolinesulfuric acid (ferron) chelates,[35] the organic dyes, camphorquinone [36] and erythrosin B[37] and metalloporphyrins.[38] The following sections describe optical oxygen sensors using metalloporphyrins and metallophthalocyanines.

4.2   Phosphorescence Quenching-based Sensors

In this section, the determination of oxygen concentrations by measuring the phosphorescence intensities of platinum octaethylporphyrin, PtOEP, in the presence of oxygen is described.

PtOEP-polymer sensing films are prepared by incorporating PtOEP in gas-permeable, solvent-impermeable polymers, such as silicone rubber, polystyrene and inorganic polymers and their optical sensing properties are determined using phosphorescence intensity measurement.

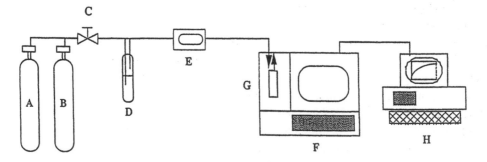

Fig. 4.1   Schematic representation of experimental set-up; A: argon cylinder, B: oxygen cylinder, C: needle valve, D: oil bubbler, E: gas flow meter, F: luminescence spectrometer, G: sample compartment, and H: personal computer. (Reproduced with permission from *Spectro. Chim. Acta A,* **54** (1998) 91, Elsevier Science)

Other characteristics of the sensing films as oxygen sensors, including reproducibility, stability and response time, are also described. To investigate sensing properties of sensing films to different oxygen concentrations, the sensor system is constructed as shown in Fig. 4.1.[39] The system consists of a gas-flow delivery part (A, B, C, D and E) and a measurement and recording part (G, F and H). Gas calibrations are determined as well-defined oxygen percentage using a gas flow meter (E) which controls the relative flow rates of nitrogen and oxygen. Luminescence spectra are obtained with a luminescence spectrometer (F) which employs a 150 W xenon-pulsed excitation source and is equipped with a laboratory-made data station (H). A gas flow meter is utilized to measure the relative flow rates of oxygen and nitrogen. The oxygen concentration is calculated by dividing the oxygen flow rate by the sum of the total flow rate of the mixed gases.

The room-temperature phosphorescence spectra of a representative sample under deoxygenated, ambient and oxygenated conditions are shown in Fig. 4.2. The corrected phosphorescence maximum is typically seen at 646 nm while the excited one is seen at 535 nm. These spectra indicate the effective quenching of room-temperature phosphorescence by oxygen. Although phosphorescence maxima do not change with different oxygen concentrations, the relative intensity decreased by 94.6% on changing from deoxygenated to oxygenated conditions, indicating the applicability of this sensor to oxygen detection. The absorption and room-temperature phosphorescence spectra of immobilized PtOEP in the polymers are almost the same as those

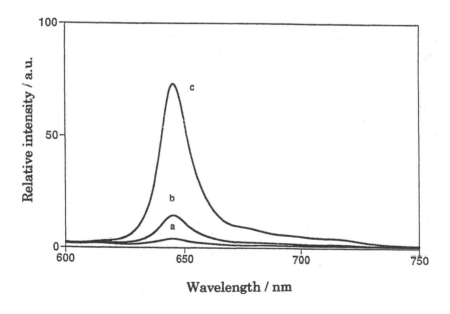

Fig. 4.2   Luminescence spectra of PtOEP-PS film under different atmospheric conditions: (a) oxygenated, (b) ambient, and (c) deoxygenated conditions. Excitation wavelength:535 nm. (Reproduced with permission from *Spectro. Chim. Acta A*, **54** (1998) 91, Elsevier Science)

found in solvents. From the above results it can be concluded that there is no electronic interaction between PtOEP and polymer, hence the polymers used can act as chemically inert matrices for PtOEP. It was also found that the PtOEP-based sensor can use light-emitting diodes (LEDs) as the visible-light excitation source, this from the result that absorption bands of the film showed a good match with the emission bands of LEDs. In a homogeneous system, the intensity from the photoexcited state of PtOEP is quenched by oxygen according to the Stern-Volmer equation:

$$I_0 / I = 1 + K_{sv} [O_2] \qquad\qquad (4\text{-}3)$$

where I_0 and I are phosphorescence intensities in the absence and in the presence of oxygen, respectively, and K_{sv} is the Stern-Volmer quenching constant. Oxygen sensing films showing downward curvature as a function of increasing oxygen concentration is observed as shown in Fig. 4.3, although they have large intensity values. However, the Stern-Volmer plots hold well at least for oxygen concentrations up to about 20%. Usually, the results of downward curvature in the Stern-Volmer plots at higher oxygen concentrations have been explained by two different theories, one being the coexistence of dynamic and static quenching processes and the other the fact that dynamic quenching is the sole quenching process in an optical oxygen sensor system with

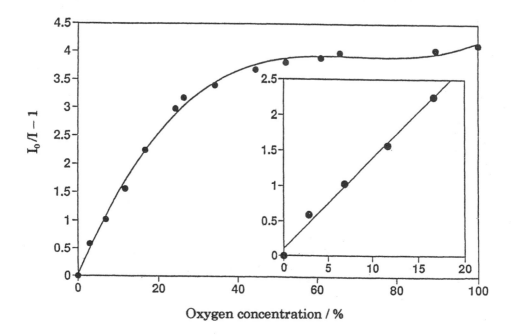

Fig. 4.3   Stern-Volmer plot of PtOEP-PS film. (Reproduced with permission from *Spectro. Chim. Acta A*, **54** (1998) 91, Elsevier Science)

differences in sites occupied by dye molecules with different oxygen accessibility. The simultaneous presence of dynamic and static quenching when the luminophore is quenched both by excited stated collisions and by ground state complex formation with the quencher, as shown in Scheme 4-1.

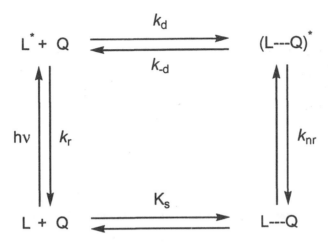

Scheme 4-1 Dynamic and static quenching process. (Reproduced with permission from *Spectro. Chim. Acta A*, **54** (1998) 91)

In the scheme L and L* are the luminophore and its photoexcited state, Q is the quencher and k_r, k_d, k_{-d} and k_{nr} are the rate constants for radiative decay, diffusion, back diffusion and non-radiative quenching, respectively. The relative phosphorescence intensity (I_0 / I) is given by the product of both dynamic and static quenching.[40] Hence

$$I_0 / I = (1 + K_D[Q]) (1 + K_S[Q]) = 1 + K_1 [Q] + K_2[Q]^2 \qquad (4\text{-}4)$$
$$I_0 / I = (1 + K_D[Q]) \exp(K_T[Q]) \qquad (4\text{-}5)$$
$$K_T = (N_a V/1000)[4pP^2(D\tau)^{1/2}] \qquad (4\text{-}6)$$

where, K_D and K_S are the dynamic and static quenching constants, respectively and $K_1 = K_D + K_S$ and $K_2 = K_D K_S$. If there is a transient component present in dynamic quenching,[40] then K_T is the transient quenching constant, N_a is the Avogadro number and V and R are volume and radius of the transient quenching sphere, and D and T are viscosity of the medium and the lifetime of the luminophore, respectively. For weak quenching,

$$I_0 / I = (1 + K_D[Q]) (1 + K_T[Q]) = 1 + K_1 [Q] + K_2[Q]^2 \qquad (4\text{-}7)$$

Thus overall dynamic quenching accompanied by either static quenching or a transient component in dynamic quenching can result in the quadratic dependence of [Q]. However, the quadratic form from eq.(4-4) or (4-7) will result in upperward curvature because K_2 is positive. On the other hand, the experimental results show downward curvature (Fig. 4.4). Thus, the latter seems to be more reasonable, taking the heterogeneous polymer matrix of these sensing systems into consideration. The quenching behavior of photoluminescent molecules is affected by the structure of the matrix and by the local composition in which the molecule is situated. Especially in heterogeneous systems, it is believed that there are inhomogeneities in the dye environment, i.e., two or more sites with different quenching constants occupied by PtOEP molecules: quencher-easy accessible and quencher-difficult accessible sites.[41,42] Therefore, in the case of a condensed system such as polymer film, the Stern-Volmer plot deviates from linearity because of the different relative contributions which originate from different quenching sites. Hence, the dependence of I_0 / I on the oxygen concentration differs from eq. (4-3).

$$I_0 / (I_0 - I) = 1 / [\Sigma (f_i K_{qi}[O_2]) / (1 + K_{qi}[O_2])] \qquad (4\text{-}8)$$

where $I_0 = \Sigma I_{0i} / n$, $I = \Sigma I_i / n$, $f_i = I_{0i} / I_0$ and n is the number of dye molecules.

Assuming that there are k accessible (and n-k non-accessible) molecules with the same K_q, then eq. (4-8) can be expressed as

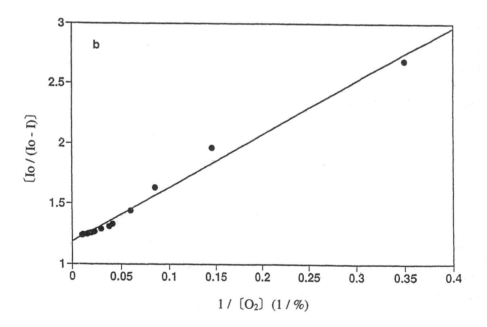

Fig. 4.4 Modified Stern-Volmer plots for the PtOEP-PS film. (Reproduced with permission from *Spectro. Chim. Acta A*, **54** (1998) 91, Elsevier Science)

$$I_0 / (I_0 - I) = 1 / f K [O_2] + 1/f \qquad\qquad (4\text{-}9)$$

where $f = \Sigma\, f_i$. This is the maximum mole fraction of dye molecules that are accessible to oxygen. If only a single class of dye molecules with the same accessibility to oxygen is present, $1/f$ in eq. (4-8) should be 1.

Figure 4.4 shows the modified Stern-Volmer plots for the PS films. In order to obtain the f value, the regression line is extrapolated to $1 / [O_2] = 0$. The value of f is 0.841 for PS-based sensing films. The value of f indicates the oxygen-quenching molar fraction of the PtOEP molecule and $f = 1$ means that all excited PtOEP molecules are quenched equally by oxygen. Sensing film with high oxygen diffusion and solubility is a good candidate for this purpose.

Generally, oxygen-sensitive dyes suffer from photodecomposition. Therefore, Stern-Volmer plots should be tried by the following method. Upon irradiation with 535 nm light, a gradual decrease in photoluminescence intensity is observed over the first 10 - 30 min. After this time a stable phosphorescence intensity is observed after a few hours of illumination. Steady-state measurements are generally performed after this illumination/conditioning period. The photostability of oxygen-sensing films is tested under continuous irradiation of UV xenon-lamp for 24 h. The phosphorescent oxygen probe, PtOEP, when immobilized in polymers, shows that 8.0 - 13% of intensity decrease is observed in the presence of oxygen. This stability is attributed to the high gas-permeable, solvent-impermeable properties of polymer matrices.

This sensor undergoes no significant response deterioration over a period of one year. After storage for one year at room temperature in the dark, the relative intensity for gaseous oxygen is about 90% of the original level. The ultimate lifetime of optical sensors depends upon the stability of the film components.

A typical example of the relative intensity changes-versus-time curve for the response at the PtOEP sensing films to change between fully oxygenated and fully de-oxygenated atmospheres is shown in Fig. 4.5. The response times (τ_{90}; 90% of total intensity change to occur) of PS-based sensing film are 35 s on going from nitrogen to oxygen and 100 s on going from oxygen to nitrogen, respectively, and 3% signal decrease is observed during the measurement. As shown in Fig. 4.5, PS-based sensing film also showed a 5% decrease during the measurement. Reproducibility test is performed for 100 min by continuous illumination under different oxygen concentrations before and after the illumination/conditioning period. Before the illumination/conditioning period, one-third of the original intensity value is obtained. However, after the illumination/conditioning period, less than 5% decrease in intensity value is observed.

Ceramic materials such as glass have been widely used as supports for optical sensors owing to their superior properties over organic polymers.[43-50] As solid supports for chemical reagents including dyes, inorganic glass sensors have distinct advantages: (1) glass materials and inorganic ceramics prepared by the sol-gel process are transparent, making them highly suitable for quantitative spectrophotometric tests,[51,52], (2) the glasses are chemically inert, photostable and thermally stable, compared with, for example, polymer matrices, making them highly suitable

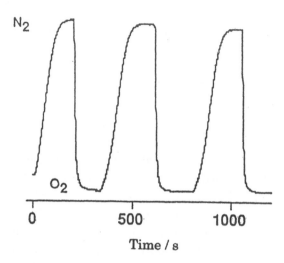

Fig. 4.5　Luminescence intensity change of PtOEP-PS film with alternate exposures to 100% nitrogen
　　　　and to 100% oxygen. (Reproduced with permission from *Spectro. Chim. Acta A*, **54** (1998) 91,
　　　　Elsevier Science)

for applications in harsh environments, and (3) the preparation of the doped glasses is technically
simple. Thus, the trapping procedure is straightforward and non-specific compared with the
covalent binding of a reagent to a solid support. The reagents are generally non-leachable, thus
offering clear advantages over reagent adsorption techniques. Finally, glass materials of various
shapes, as well as thin films, are easily prepared. This section describes the fabrication of porphyrin-
doped sol-gel glasses and the optical-oxygen sensing properties of these materials.[53] Other sensor
characteristics such as photostability and storage stability are also described. Recently, a novel
doping method has been introduced using the sol-gel polymerization process. The dopant (in this
case, PtOEP) is incorporated in the sol-gel glass at the early stages (or even before initiation) of
the polymerization step. Thus, when the dry xerogel is formed the dopants remain physically
encapsulated within the glass matrix but maintain their ability to interact with diffusing species.
An optical oxygen sensor using PtOEP-doped silica gel glasses is described below.

　　　　Absorption maxima of the PtOEP-doped silica glass show a red shift, compared with the
initial acetone solution (absorption maxima = 536, 502 and 380 nm in glass; 533, 500 and 377 nm
in acetone solution). This red shift can be explained by the lower polarity of the silica matrix,
which consists of siloxane (Si-O-Si) and silanol (Si-OH) groups.[44,54,55] Phosphorescence spectra
of the PtOEP-doped silica glass under de-oxygenated, ambient and oxygenated conditions are
shown in Fig. 4.6, indicating effective quenching of the phosphorescence intensity by oxygen.
The emission spectrum of the silica matrix shows no red (or blue) shift and no differences in peak
shape, compared with that of the acetone solution (646 nm in both instances). The emission

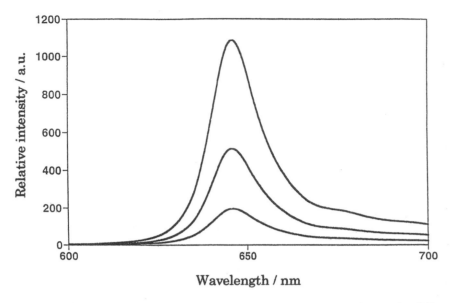

Fig. 4.6   Luminescence spectra of PtOEP-doped based-catalyzed sol-gel glass under different atmospheric conditions:(a) oxygenated, (b) ambient, and (c) deoxygenated conditions. Excitation wavelength, 535 nm. (Reproduced with permission from *Spectro. Chim. Acta A*, **54** (1998) 91, Elsevier Science)

maxima intensities increase strongly on going from the oxygenated to the ambient and deoxygenated conditions. Hence, the sol-gel process can provide a good matrix with no interactions between PtOEP and silica during the optical sensing of oxygen.

The phosphorescence intensity from the excited state of PtOEP immobilized in the silica matrix is quenched by oxygen according to the Stern-Volmer equation. In order to minimize photodecomposition of the samples when plotting the Stern-Volmer curves, the samples are kept in the dark except during scanning. Fig. 4.7 shows the sensitivity (I_0 / I) and linearity of the sol-gel sensors prepared at different drying temperatures. The sensors that are prepared by drying at room temperature (sample 1) and 150°C (sample **2**) exhibited much higher sensitivities than the sensor prepared by drying at 180°C (sample 3). These results indicate that a "sintering step" is not necessary. The room temperature-dried sample shows good sensitivity to oxygen. The linearity is also different. Sample 1 shows an upward curvature, whereas samples 2 and 3 show a downward curvature. All three samples are non-linear, although the best sensitivity is obtained with sample 2. The effect of the drying temperature on the sensitivity and linearity of the sensors described here may be explained by several factors or by a combination of factors. (1) Oxygen permeability and diffusion through the sol-gel matrix. Oxygen permeability and transport through the sol-gel matrix will greatly affect the sensitivity and linearity as well as the response time of the sensor. During the aging and drying process, changes in the physical properties of the gel occur. Aging

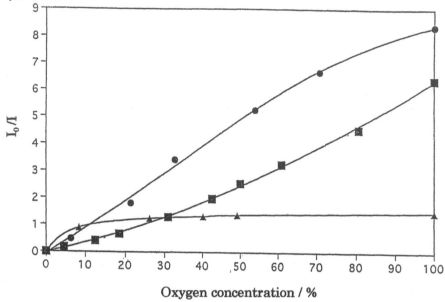

Fig. 4.7   Stern-Volmer plots of relative phosphorescence intensity for PtOEP-doped based-catalyzed
sol-gel glasses with different drying temperatures: top (sample 2), middle (sample 1) and
bottom (sample 3). (Reproduced with permission from *Spectro. Chim. Acta A*, **54** (1998)
91, Elsevier Science)

and thermal treatment result in an increase in pore size. Since the pores can act as an oxygen
channel to dye molecules immobilized in the sol-gel matrix, the increase in pore size results in
increased molar fractions of oxygen-quenchable dye, hence the resulting sensitivity and linearity
of the sensor are modified. This explains why the linearity and sensitivity of sample 2 are superior
to those of sample 1. However, this explanation cannot account for the low sensitivity and marked
deviation from linearity of sample 3. (2) Dye molecule distribution in the sol-gel matrix. If the
distribution of the dye in the sol-gel matrix changes with drying temperature by the migration of
the dye molecules into different sites within the matrix and/or by partial diminishing, this affects
the sensitivity and linearity of the sensor. This suggests that the micro-environment of the sensor
materials varies with the drying temperature. In fact, the optical transparency and color of the
samples are found to change with increasing temperature. The completely transparent pink sample
1 (dried at room temperature) keeps its transparency up to 150°C, but become slightly opaque
and yellow when dried at 180°C. This indicates a change in the micro-environment of the doped
sol-gel. The quenching behavior of photoluminescent molecules is affected by the structure of the
matrix and by the local composition in which the molecule is situated. In heterogeneous systems,
it is believed that there are two or more sites with different quenching constants: quencher-easy
accessible and quencher-difficult accessible sites. Therefore, the Stern-Volmer plot deviates from
linearity because of the different relative contributions, which originate from different quenching
sites.

Figure 4.8 demonstrates the typical dynamic response of the sensor when switching
between fully oxygenated and fully de-oxygenated atmospheres. The response time of the sensor

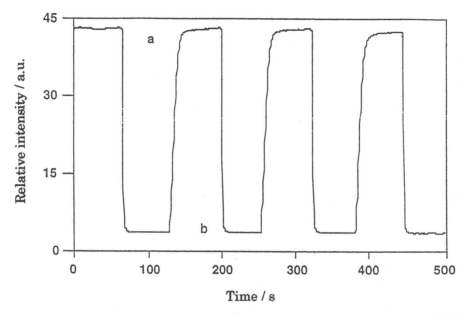

Fig. 4.8   Response time, relative intensity change and reproducibility for sample 1 on switching between 100% nitrogen (a) and 100% oxygen (b). (Reproduced with permission from *Spectro. Chim. Acta A*, **54** (1998) 91, Elsevier Science)

is about 5 s on going from nitrogen to oxygen and 10 s on going from oxygen to nitrogen. The times required for achieving 90% of the ultimate response, t (90%) for both cases are less than 5 s. These are very short response times and are similar to those reported by McEvoy et al. for their optical sensor.[56] The fast response obviously comes from the microporous texture of the sol-gel matrix. Such fast response times are not common among polymer matrix-derived optical sensors. As is also shown in Fig. 4.9, stable and reproducible signals are obtained with the sensor. The fact that the inorganic matrix prepared by the sol-gel method is responsible for the enhanced stability compared with polymer matrix and other sensor production methods is verified by the stability test (*vide infra*).

The photostability is tested by placing the PtOEP-doped silica glass in front of a UV xenon lamp, with continuous irradiation under ambient conditions. Fig. 4.9 shows the photostability of the sensor (sample 1) under different atmospheric conditions. The PtOEP-doped silica glass exhibits improved photostability, compared with the PtOEP-adsorbed silica sample [Figs. 4.9 (b) and (c)]. Although a 10% decrease in intensity is observed in a fully oxygenated environment, no decrease in intensity occurs in inert and ambient atmospheres even after a measurement time of 100 min. It should be noted that it is necessary to reverse to an inert atmosphere after carrying out measurements under oxygenated conditions. With continuous illumination for 24 h in an ambient atmosphere, the decrease in the original luminescence intensity is < 15%. These results are comparable to those of a previous photobleaching test by Papkovsky et al., who reported that

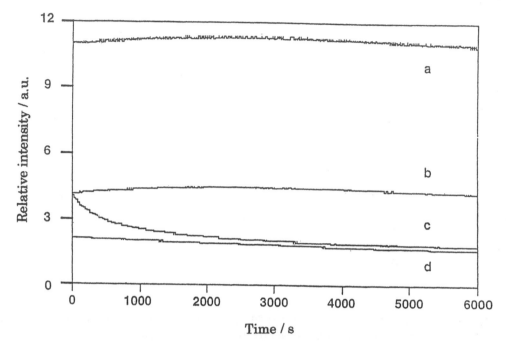

Fig. 4.9 Photostability of PtOEP-doped based-catalyzed sol-gel glass (sample 1) in inert (a), ambient (b) and oxygenated atmosphere (d) and adsorbed silica sample (c). (b) and (c) are normalized to allow a comparison of the decrease in intensity in each instance. (Reproduced with permission from *Spectro. Chim. Acta A*, **54** (1998) 91, Elsevier Science)

a porphyrin ketone/polymer sensor showed 88% recovery of the initial intensity under similar conditions.[57] The PtOEP-doped sol-gel sensor exhibited no significant change in response, intensity or sensitivity after five months in the absence of light.

4.3 Fluorescence Quenching-based Sensors Using Metallophthalocyanine

A number of luminophores have been used as oxygen sensing probes for optical oxygen sensors.[58-65] Many optical oxygen sensors are composed of organic dyes such as polycyclic aromatic hydrocarbons (pyrene and its derivatives, quinoline and phenanthrene) immobilized in oxygen permeable polymer.[58,59] Polycyclic aromatic hydrocarbons have strong luminescence with long fluorescence lifetimes, e.g. 200 ns, making them very amenable to quenching by oxygen. However, these compounds only absorb light in the ultra-visible and near-visible range and suffer from the disadvantage of small Stork's shift. Thus, these compounds require complicated and expensive optical instrumentation for the separation of the excitation light from the luminescence signal. On the other hand, metal porphyrins and phthalocyanines have absorption maxima in the visible region.[66-68] As the absorption maxima of metal phthalocyanines are around 600 nm, a convenient light source such as a He-Ne laser can be used for the excitation of phthalocyanines. Among the metallophthalocyanines, aluminum phthalocyanine has a fairly long fluorescence lifetime.[69-72] Thus, aluminum phthalocyanines are attractive candidates for a new optical oxygen sensing system using fluorescence quenching.

In this section a new optical oxygen sensing based on fluorescence quenching of aluminum phthalocyanine, aluminum 2,9,16,23-tetraphenoxy-29H,31H-phthalocyanine hydroxide (AlPc(OH); the structure is shown in Fig. 4.10) by oxygen is described. To investigate the sensing properties of sensing films to different oxygen concentrations, the sensor system is constructed as shown in Fig. 4.2. The film shows strong fluorescence at 706 nm when excited at 606 nm, attributed to the Q-bands of AlPc(OH), as shown in Fig. 4.11 The fluorescence intensity of the film depends on the oxygen concentration as shown in Fig. 4.12. This result indicates that the fluorescence of AlPc(OH) in PS film is quenched by oxygen and that this film can be used as an optical oxygen-sensing device based on fluorescence quenching by oxygen. Fig. 4.13 shows the plot of the fluorescence of AlPc(OH) as a function of oxygen concentration (Stern-Volmer plot, $I_0 / I = 1 + K_{SV} [O_2]$; where I_0 and I are fluorescence intensities in the absence and in the presence of oxygen, respectively; K_{SV} is the Stern-Volmer quenching constant). The plot exhibits considerable linearity at lower oxygen concentrations (~10%). Fig. 4.14 shows the typical dynamic response of the sensor when switching between fully oxygenated and fully deoxygenated atmospheres. The response times of the sensor are 28 s on going from argon to oxygen and 33 s on going from oxygen to argon. This result indicates that this sensing film possesses a fast response time. The signal changes are fully reversible

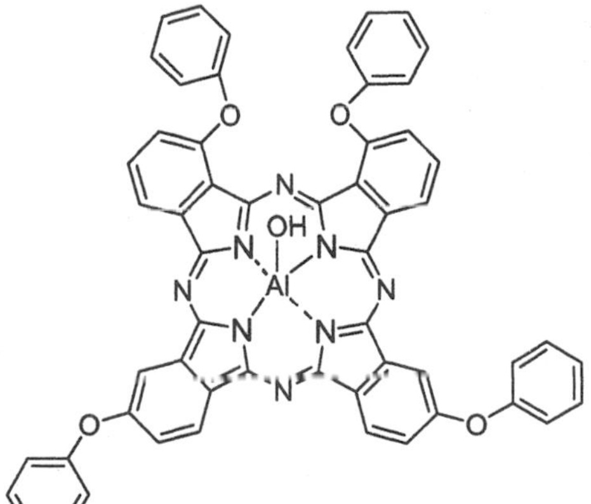

Fig. 4.10 Structure of aluminum 2,9,16,23-tetraphenoxy-29H, 31H-phthalocyanine (AlPc(OH)).

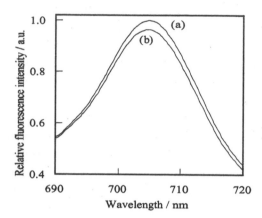

Fig. 4.11 Fluorescence spectrum change of AlPc(OH)-PS film. (a) deoxygenated and (b) oxygenated condition, respectively. Excitation wavelength is 606 nm.

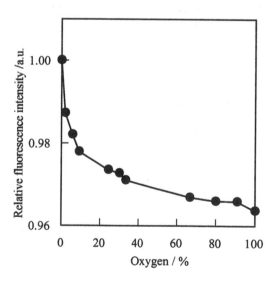

Fig. 4.12 Fluorescence intensity change of AlPc(OH)-PS film by oxygen. Excitation and monitored wavelength were 606 and 706 nm, respectively.

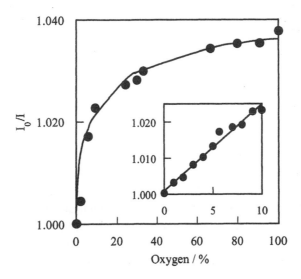

Fig. 4.13 Stern-Volmer plot for AlPc(OH)-PS film. Excitation and monitored wavelength were 606 and 706 nm, respectively. The inset is Stern-Volmer plot in the low oxygen concentration region (0-10%).

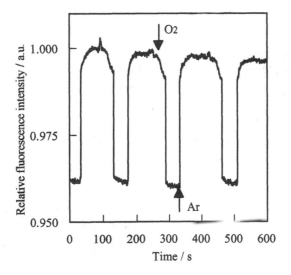

Fig. 4.14 Operational stability and response time of the AlPc(OH)-PS film. Excitation and monitored wavelength were 606 and 706 nm, respectively.

and hysterisis is not observed. The photostability is tested by AlPc(OH)-PS film in front of a visible light lamp with continuous irradiation under ambient conditions. As no decomposition is observed for 4 h irradiation, this film is stable against irradiation.

4.4 Luminescence Lifetime-based Sensors

For optical sensing, the luminescence intensity changes by oxygen have been used. Although intensity measurements are simple, the intensity depends on the thickness of the sensing film, scattering light from background and the concentration of dye in film. On the other hand, the lifetime of the photoexcited triplet state involves only the light emission following a flash of excitation light and is not dependent on the intensity of the emitted light. The concept of using lifetime measurement of the photoexcited triplet state of dye has proven to be especially attractive for optical oxygen sensing, and the method is more suitable than steady state luminescence quenching methods.[73,74]

The principle of luminescence lifetime measurement and oxygen sensing application is shown in Fig. 4.15. Luminescence lifetime measurements are carried out using Nd-YAG laser (Spectra Physics, pulse width 10 ns) at room temperature. The excitation wavelength is the Q band of PtOEP. The light reflected by oxygen-sensing film is transmitted by a quartz optical fiber to a monochromator and photomultiplier. The decay curves are obtained by averaging 256 single shots. Decay curves were analyzed without deconvolution by fitting with the following equation:

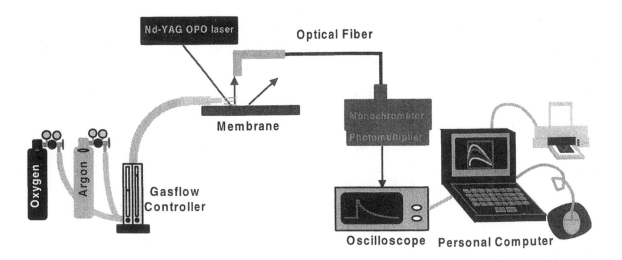

Fig. 4.15 Oxygen sensing system for measuring of luminescence lifetime of PtOEP-PS film.

$$R(t) = \Sigma A_n \exp(-\tau / \tau_n) \qquad (4\text{-}10)$$

where $R(t)$ is time-resolved reflectance and t is the lifetime of the photoexcited triplet state. A_n is the fractional contribution to each lifetime component. The calculated values of $R_c(t)$ are then compared to the experimental $R(t)$ values and the goodness-of-fit is determined by minimization of the χ^2 function

$$\chi^2 = \Sigma w_n [R(t) - R_c(t)]^2 \qquad (4\text{-}11)$$

where w_n is a statistical weighting factor that accounts for the uncertainty in each $R(t)$ value. When w_n accurately describes the uncertainty in $R(t)$, χ^2 and random distribution of residuals $[R_c(t) - R(t)]$ are equal to 1.0 and zero, respectively.

Figure 4.16 shows typical decays of the luminescence of PtOEP-PS film under argon (1) and oxygen (2) monitored at the emission peak of PtOEP. The decays consist of two components (faster and slower lifetimes) in the absence and presence of oxygen. The faster decay of the luminescence of platinum porphyrin under oxygen, compared to that under argon indicates that the quenching of luminescence of PtOEP by oxygen occurrs. Fig. 4.17 (1) shows the luminescence lifetimes of PtOEP with oxygen concentration. The lifetimes of platinum porphyrin are calculated by a fitting with the eq. (4-10). In both the faster and slower components, the lifetime decreases with increasing oxygen concentration.

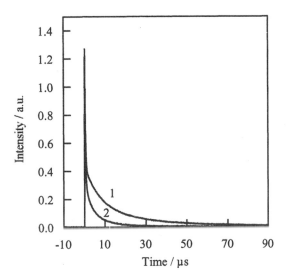

Fig. 4.16 Typical luminescence decay of PtOEP-PS film monitored at 646 nm. (1) argon saturated; (2) oxygen saturated. Excitation wavelength: 535 nm.

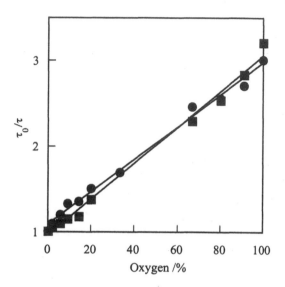

Fig. 4.17 Luminescence lifetime of PtOEP-PS film with oxygen concentrations. ■: faster component, ●: shorter component.

Figure 4.17 shows a Stern-Volmer plot ($\tau_0/\tau - 1 = K_{sv}[O_2]$) where τ_0 and τ are lifetimes in the absence and presence of oxygen, respectively. In both components, a Stern-Volmer plot of PtOEP-PS films exhibits linearity. However, K_{sv} strongly depends on the τ_0 and quenching rate constant k_q ($K_{sv} = k_q \tau_0$). The rate constant of quenching in the faster component is larger than that of the slower component, indicating that two different oxygen-accessible sites exist in PtOEP-PS film. The faster and slower components contribute to the oxygen-accessible site on the surface and bulk of PtOEP film, respectively. In the fractional contribution to each lifetime component, the contribution of the faster component is larger than that of the slower component, indicating that the sensing site on the surface is important for optical sensing.

4.5 Oxygen Sensing System by Triplet-Triplet Absorption of Porphyrins

An optical oxygen sensing system based on the phosphorescence of various platinum porphyrin-doped polymers or sol-gel glass has been described above. However, most organic compounds have no luminescence at room temperature so the number of oxygen-quenchable compounds is extremely limited. Triplet-triplet (T_1-T_n) absorption on a flash photolysis set-up allows us to decide the excited triplet lifetime of such non-luminescent compounds.[76-78] Analysis of the quenching of the photoexcited triplet state of a compound is based on the decrease in lifetime in the presence of oxygen. If the quenching of the photoexcited triplet state is entirely diffusional, the lifetime is related to the oxygen concentration by the Stern-Volmer equation

$$\tau_0 / \tau = 1 + K_{SV}[O_2] \tag{4-12a}$$

$$K_{SV} = k_2 \tau_0 \tag{4-12b}$$

where t is the lifetime of the photoexcited triplet state. The subscript "$_0$" denotes the value in the absence of oxygen. K_{SV} is the Stern-Volmer quenching constant, and k_2 is the bimolecular quenching constant. Plots of τ_0 / τ versus $[O_2]$ will be linear with identical slopes of K_{SV}. The above method extends the number of indicators available for the oxygen sensor. In this section a new system using T_1-T_n absorption quenching of zinc tetraphenylporphyrin-doped polystyrene membrane (ZnTPP-PS) for measuring oxygen concentration is described.[78]

Upon absorption of light a molecule is raised from the singlet ground state to an excited singlet state which by intersystem crossing may convert to the triplet manifold. By internal conversion the lowest triplet state (T_1) can be produced in a relatively high concentration. Therefore, it is not difficult to measure absorption spectra which are due to the spin-allowed transitions from T_1 to higher triplet states T_n. Because the T_1 to S_0 transitions are forbidden, the lifetimes of the triplet can be orders of magnitude longer than those of the singlet excited states. Typically, fluorescence lifetimes of polycyclic aromatic molecules are less than 100 ns, while the lifetime of

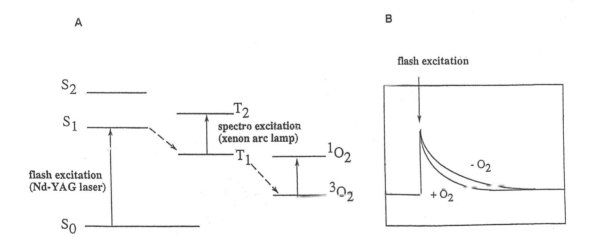

Fig. 4.18 Schematic diagrams of laser flash photolysis (A) and time profiles of porphyrin with and without oxygen (B).

the triplet state can be as long as several seconds, particularly in rigid media. In fluid solutions, the triplet lifetime can range from ten nanoseconds to hundreds of milliseconds, depending on the experimental conditions. The principle of laser flash photolysis and oxygen sensing application is shown in Fig. 4.18.[78] As shown in the figure, in flash photolysis techniques, a strong xenon or laser lamp is used to generate T_1 states via S_0 to S_1 absorption, followed by intersystem crossing from S_1 to T_1. The triplets are detected via the transient T_1 to T_2 absorption, which is monitored by means of a second, weaker light source. This method is particularly useful when the lifetime of the triplet is short (nanosecond range). The sensing system for measuring oxygen is shown in Fig. 4.19. Laser flash photolysis is carried out using OPO-Nd-YAG laser with pulse width of 10 ns at room temperature. A xenon arc lamp as a monitoring light beam is coupled into one end of an optical fiber. The light reflected by the oxygen-sensing membrane is transmitted by the same fiber to a monochromator and photomultiplier. Different oxygen standards (in the range 0-100%) in a gas stream are produced by controlling the flow rates of oxygen and argon gases entering a mixing chamber. The oxygen concentration is calculated by dividing the oxygen flow rate by the sum of the total flow rate of mixed gases. The total pressure is maintained at 760 Torr.

Figure 4.20 shows the time profiles of the triplet state of ZnTPP (monitored at 470 nm) under argon (1) and oxygen (2) after laser flash photolysis. The decays in the absence and presence

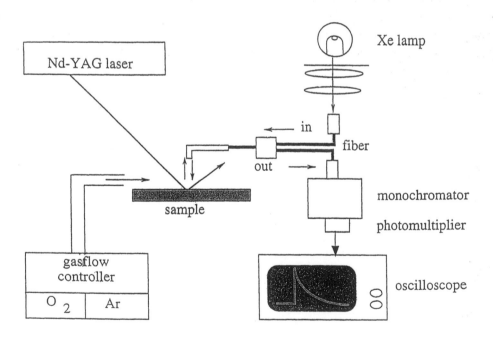

Fig. 4.19 Oxygen sensing system for measuring T_1-T_n absorption of porphyrin sensing film. (Reproduced with permission from *Chem. Lett.,* (**1998**) 61)

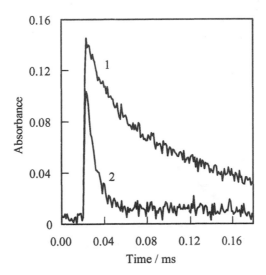

Fig. 4.20 Decay curve of the photoexcited triplet state of ZnTPP-PS film under 0% (1) and 100% (2) oxygen. (Reproduced with permission from *Chem. Lett.,* **(1998)** 61)

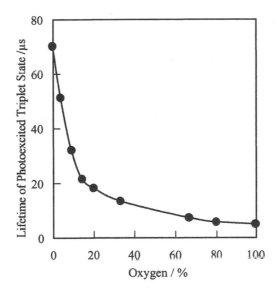

Fig. 4.21 Lifetime change of the photoexcited triplet state of ZnTPP-PS film under various oxygen concentrations. (Reproduced with permission from *Chem. Lett.,***(1998)** 61)

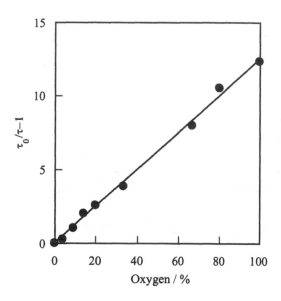

Fig. 4.22 Stern-Volmer plot for ZnTPP-PS film. (Reproduced with permission from *Chem. Lett.,* (**1998**) 61)

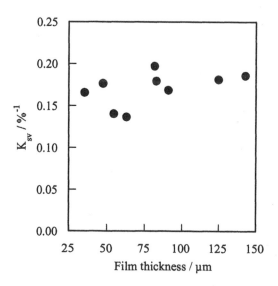

Fig. 4.23 Effect of film thickness on Stern-Volmer coefficient.

of oxygen obey first-order kinetics. The faster decay under oxygen, compared to that under argon, indicates that the excited triplet state of ZnTPP is quenched by oxygen. This result indicates that the ZnTPP membrane can be used as material for the oxygen sensor. Under the reaction conditions the decays of the T_1-T_n absorption of ZnTPP obey first-order kinetics. Fig. 4.21 shows the photoexcited triplet lifetimes of ZnTPP with oxygen concentration. The triplet lifetimes of ZnTPP are calculated by the first-order plot. The lifetime decreases with increasing oxygen concentration. Fig. 4.22 shows a Stern-Volmer plot ($\tau_0/\tau - 1 = K_{sv}$ [O_2], where τ_0 and τ are triplet lifetimes in the absence and presence of oxygen, respectively. [O_2] is the concentration of oxygen and K_{sv} is the Stern-Volmer quenching constant.). A Stern-Volmer plot of ZnTPP exhibits considerable linearity (r = 0.988). The sensing properties are strongly affected by the thickness of the ZnTPP-PS membrane. In theory, a thinner membrane needs less time for the oxygen inside of the membrane to reach equilibrium with the outside environment. Fig. 4.23 shows the relationship between the Stern-Volmer constant and membrane thickness. The Stern-Volmer constant (K_{sv}) seems to be little affected by differences in membrane thickness.

4.6   Oxygen Monitoring in Living Cells by Palladium Porphyrin

Accurate and rapid measurement of oxygen concentration, particularly low oxygen concentrations in biological samples has been the goal of many research workers. By far the most successful method has involved use of the oxygen electrode.[79] However, it is limited by the stability of the electrode surface and by instability in the oxygen diffusion barrier (because it measures the rate of diffusion of oxygen to the cathode). The oxygen-dependent quenching of phosphorescence has proven to be a powerful method for measuring oxygen pressure in biological samples.[80] The method has a rapid response time (ms) and it is possible to measure oxygen pressure throughout the physiologically important range (760 Torr down to 10^{-2} Torr). One of the most promising applications is the measurement of oxygen pressure in living cells. Recently, an optical method has been reported for measuring oxygen pressure in normal tissues,[81] tumors[82] and living cells.[83] The method is based on the oxygen-dependent quenching of phosphorescence of porphyrins or luminescence of ruthenium tris-bipyridine (Ru(bpy)$_3^{2+}$) and yields two-dimensional maps of intravascular oxygen pressure. This method has the advantage that the oxygen pressure information is related only to the time-dependent features of the phosphorescence signal and not to the intensity of emission. Therefore, the information of interest is neither concentration- nor light intensity-dependent. Oxygen-dependent quenching of phosphorescence provides very sensitive measurement of oxygen pressure in the environment of phosphorescent molecules. Palladium and platinum porphyrins especially display strong room-temperature phosphorescence with high quantum yield and long lifetime.[84] Therefore, these compounds are useful in optical oxygen sensing. Among porphyrins, palladium *meso*-tetra(4-carboxyphenyl)porphyrin (PdTCPP) is suitable for experiments using living cells because TCPP is unaffected by changes in pH and ionic composition of the medium.[85)] Oxygen monitoring in living cells using a phosphorescence signal from PdTCPP is described below. MH134 and HeLa cells are used mainly for photometric measurements and fluorescence imaging, respectively. To elucidate localization and distribution of PdTCPP in the cells, fluorescence images are obtained by fluorescence microscope. Samples are excited at 404 and 435 nm through a 400-440-nm bandpass filter and imaged through a 475-nm cut-off filter. MH134 cells (3.0 ml) are extracted from tumor-bearing mice. After centrifugation with PBS, they are incubated with 10^{-4} mol dm^{-3} of PdTCPP for 2 h. Spectrophotometric measurements are conducted after washing with PBS twice and centrifugation again. In the case of HeLa cells, 2×10^5 cells are subcultured into 35-mm Petri dishes containing 2.0 ml of growth medium. After incubation with 1.0×10^{-4} mol dm^{-3} of PdTCPP for 2 h, the dye solution is removed and then the cells are washed with fluorescence microscopic observations.

Oxygen monitoring in living cells is conducted using a phosphorescence intensity method. Bubbling of high quality nitrogen into cell-containing samples for more than 20 min is conducted for measurement under anaerobic conditions. The luminescence spectrum of PdTCPP-stained MH134 is similar to that of HeLa cells. Phosphorescence intensity is measured in MH134 cells under different conditions, anoxia and oxygenation. Anoxia and maximal oxygenation of cells are established by flow of nitrogen and oxygen gases into the cells, respectively. The intensity decreases by approximately one fourth when cells are fully oxygenated. These results demonstrate that phosphorescence is an effective indicator of cell oxygenation levels.

Phosphorescence intensity detected by spectrophotometer shows only the average value of probe molecules inhomogeneously distributed in the cell. In order to evaluate the distribution of the probe (and/or oxygen) in the cell, fluorescence microscopy is used. As shown in Fig. 4.24 (see the color frontispiece) the administration of PdTCPP resulted in good localization in the tumor mass, which exhibits intense emission from all illuminated cells. Figs. 4.24 (a) and (b) show the images of PdTCPP-stained HeLa cells taken using phase contrast mode and fluorescence mode, respectively. Tumor cells are significantly more luminescent than the surrounding cells, and PdTCPP is homogeneously distributed throughout the cells. This indicates that the oxygen concentration is constant throughout the cells. The images show non-uniform luminescence intensity across the cell. This can be due to either to (a) an inhomogeneous partitioning of the probe across the cell or to (b) differences in oxygen concentration within the cell. The former possibility is ruled out by the phosphorescence imaging. A recent work on fluorescence lifetime imaging of oxygen in living cell using ruthenium tris-bipyridine (Ru(bpy)$_3^{2+}$) also reported that the heterogeneities in the fluorescence intensity image can be attributed to different partitioning of the probe and not to differences in oxygen concentration within the cell.[83)]

As an alternative for luminescence intensity measurements, the lifetime method is also useful to monitor oxygen in living cells. Although intensity measurements are simple, the most common difficulty with them is that (a) they are influenced by other chromophores in the system which absorb either the excitation or emitted light and (b) the intensity of phosphorescence cannot be distinguished from fluorescence emitted in the detected wavelength range. Neither of these processes affects phosphorescence lifetime measurements. Lifetime measurements involve only the rate of decrease of the light emission following a flash of excitation light and are not dependent on the absolute intensity of the emitted light. In this context, lifetime measurements are superior to the intensity method. In many biological systems, oxygen is the only quencher present in significant concentrations.[81] In this case, the oxygen pressure in eq. (4-13) can be replaced with the oxygen concentration.

$$pO_2 = 1/K_q(1/\tau - 1/\tau_0) \tag{4-13}$$

The most important characteristic of the lifetime method is that the calibration is dependent only on the phosphor and its molecular environment. When the environmental parameters are held constant, the calibration is absolute; once the calibration constants are determined they can be generally applied. The lifetime of MH134 (1.0×10^6 cell dm^{-3}) with PdTCPP (1.0×10^{-4} mol dm^{-3}) is 362 μs. This value corresponds to 5.2 Torr based on previous calibration data (for PdTCPP = 710 μs and K_q = 260 $torr^{-1} s^{-1}$ at 23°C, pH 7.4). This result indicates severe hypoxia inside MH134 cells.

4.7   Application of the Optical Oxygen Sensor to the Aerodynamics Field

As for conventional luminescence intensity measurement, it has severe problems concerning calibration procedures and long-term reliability. The fluorescent intensity can vary due to excitation intensity, light losses in fibers and photobleaching of probe, as well as changes in light scattering or absorption characteristics of the sample. Thus, while fluorescent intensity methods may yield almost perfect results in an optically clean environment, their use in most sensing applications such as blood or microbial cultures is severely limited. Especially, for the application of pressure-sensitive coating (PSC) in the aerodynamic field e.g. for wind-off, wind-on testing and for the in-flight environment, the luminescence value must be measured for every point on the model because illumination varies over the different surfaces of the model and changes with model movement. These nonintrusive global pressure measurements cannot be easily accomplished using a laser-induced intensity technique that scans only one point or line. Recently, a charge-coupled device (CCD) camera scanning technique has been developed to detect and image oxygen pressure and/ or oxygen pressure distribution on the surface of interest.[86-88] Such pressure mapping techniques using photoluminescent quenching offer a potential savings in time and expense for wind-tunnel testing over conventional transducer.

The modified intensity measurement using CCD camera technique (video capturing method) is employed to measure oxygen pressure on a specific surface area. This promising method of oxygen detection involves optical measurements of phosphorescence quenching of platinum octaethylporphyrin (PtOEP), a strongly phosphorescent dye, by oxygen. This method proved to be suitable for the entire range of aerodynamic and environmental oxygen pressure. The objectives are to develop PSC materials and their applications for luminescent pressure measurements of high reliability for aerodynamic facilities.

Oxygen pressure measurements using PSC samples are based on the quenching phenomenon by oxygen, an active quencher of phosphorescence. In contrast to fluorescence, phosphorescence is a delayed emission, typically with longer a lifetime (from 10^{-3} s to several seconds). From the standpoint of optical sensing application, the most interesting fact is that the excited singlet and triplet states can be deactivated by quenching processes. The approximation of the luminescence intensity of the PSC is modeled by the Stern-Volmer equation.[89]

$$I_0 / I = 1 + KP \tag{4-14}$$

where P is oxygen pressure, I_0 and I are the luminescence intensities in the absence and presence of oxygen, respectively, and K is the Stern-Volmer constant. However, this relationship is generally not practical for measuring I_0 in the wind tunnel environment, since the tunnel would have to be pumped down to a vacuum. Instead of trying to achieve zero oxygen conditions the intensity of luminescence at wind off is used as the reference intensity (I (ref)), and the pressure at wind off is considered the reference pressure, P (ref). In practice, this is usually the local barometric pressure. In terms of the Stern-Volmer equation this takes the form of the ratio of the Stern-Volmer for two different pressures. That is,

$$I_0 / I_1 = (1 + KP_1) : \text{at wind off} \tag{4-15}$$
$$I_0 / I_2 = (1 + KP_2) : \text{at wind on} \tag{4-16}$$
$$I_1 / I_2 = (1 + KP_2) / (1 + KP_1) = 1 / (1 + KP_1) + KP_2 / (1 + KP_1) \tag{4-17}$$

K and P_1 are constants. Thus eq. (4-17) may be expressed as

$$I_1 / I_2 = A + BP_2 \tag{4-18}$$

where A = [1 / (1 + KP_1)] and B = [K / (1 + $K P_1$)]. Since luminescence intensity depends on illumination intensity, values for I_1 are determined for each point on the wind tunnel model at each angle of attack. The values for A and B are then determined from the plot of I_1 / I_2 versus pressure. Equation (4-18) is the fundamental equation (a modified Stern-Volmer relation) in determining oxygen pressure by PSC.

The experimental set-up used to characterize PSCs is shown in Fig. 4.25.[90,91] A PSC sample is placed inside a vacuum chamber with an optical access window. The pressure inside the chamber can be controlled and measured by a pressure transducer. The temperature on the coating

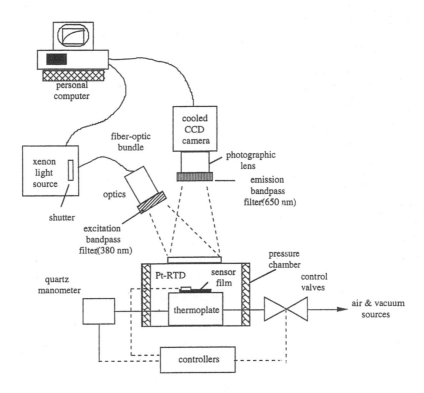

Fig. 4.25 Block diagrams of the assembly for an oxygen pressure measurement. (Reproduced with permission from *Anal. Sci.*, **13** (1997) 181)

sample is controlled by platinum RTD and monitored by a thermometer. A highly stable xenon lamp is used as an excitation source and fiber optic bundle is used to produce a spatially uniform irradiation beam. Visible light filtered by 380 ± 50 nm bandpass filter pass into the chamber and excites the luminescent sample. The luminescence emission is collected with a lens, filtered by 650 ± 20 nm bandpass filter, focused onto a photodetector and imaged onto cooled CCD camera. The luminescence signal is sampled by a personal computer when the pressure is varied. Thus, the relation between the luminescence intensity and pressure can be obtained at constant temperature. The calibration data can be fitted by the Stern-Volmer equation. The shooting condition is a camera shutter opening of F16 and exposure time of 150 ms. The number of total pixels is 512 x 512.

The modified Stern-Volmer quenching plots of several different polymer-type PSCs are shown in Fig. 4.26. Plots deviate somewhat from linearity and eq. 4.18 is poorly fitted except for GP-197 coated on aluminum plate sample. Among PSCs with the same concentration, the GP-197 sample also showed the highest oxygen quenching efficiency, good linearity and fast response time, probably due to high oxygen permeability. A and B in eq. 4.18 are 0.335 and 0.679, respectively, with a correlation factor of $(R^2) > 0.997$, and the sum of A and B is nearly 1. Based

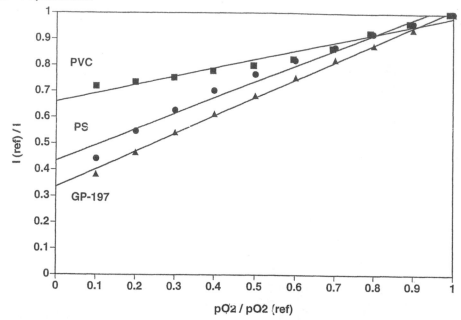

Fig. 4.26 Modified Stern-Volmer plots for the PSC sensor system. Unless indicated otherwise, I(ref) represents intensity at 100 kPa, pO$_2$(ref).   (Reproduced with permission from *Anal. Sci.,* **13** (1997) 181)

on the results of Fig. 4.26, further O$_2$ sensing performance is concentrated on the GP-197 sample.

Photostability tests are carried out by illuminating continuously for 1 h, as shown in Fig. 4.27. After illumination, PSCs showed about 20% photobleaching compared with that of nonilluminated samples, and their pressure sensitivity also deteriorated slightly. Although the mechanism of the photobleaching or photodegradation of organic dye is currently not well understood, several attempts have been made to increase the photostability of the dye immobilized in PSC by matrix modification, the modification of existing dyes. The sensing properties of PSC such as pressure sensitivity are also affected by storage time during which PtOEP is gradually photobleached. After storage for more than one year under ambient conditions, PSCs show approximately 10% decrease in luminescence intensity, compared with the original values. These results imply that calibration should be carried out before use again.

Slight deviations from linearity in the modified Stern-Volmer plots are observed in almost all cases. These results can be interpreted by the fact that there are two sites with different quenching constants. One is an oxygen-easy accessible and the other is an oxygen-difficult accessible site. The latter is assumed to be distributed in the same sensing film that is little affected by oxygen. Hence, luminescence from these sites cannot contribute to the total luminescence of dye. Demas and co-workers have proposed this concept, called microheterogeneity, evidenced in part by plotting the excited lifetimes of ruthenium complex dye immobilized in silicone.[92] When the noncontributable luminescence intensity of the sensing layer is eliminated from the total luminescence intensity, the pressure sensitivity data showed enhanced linearity. From these results,

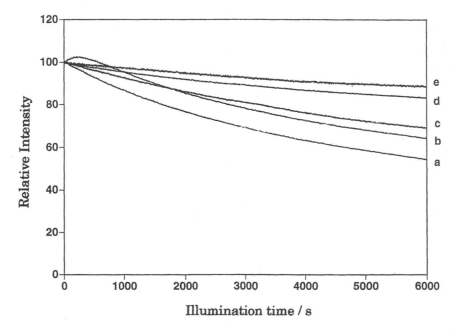

Fig. 4.27 Photostability tests are carried out under continuous illumination for 1 h. (a) PVC, (b) PS and (c) GP-197. (Reproduced with permission from *Anal. Sci.,* **13** (1997) 181)

it is more reasonable to assume that there are two types of sites with different quenching constants, i.e., oxygen-easy accessible and oxygen-difficult accessible sites in the same coating and noncontributable luminescence should be removed when analyzing oxygen pressure data quantitatively.

Pressure sensitive paint (PSP) is applied to a large transonic wind tunnel testing of a rocket fairing model at Mach number 0.9 (*ca.* 1121 km h^{-1} at room temperature).[93] A schematic drawing of the experimental setup is shown in Fig. 4.28. The experiment is conducted in a 2.0-m transonic wind tunnel. This is a closed circuit type wind tunnel and the test section is 2.0 m^2, with a slotted ceiling and floor. The uniform flow Mach number is 0.1 to 1.4 and the total pressure is 50 to 150 kPa. The PSP used in this experiment consists of PtOEP and GP197. A highly stable xenon lamp is used as the excitation source and a fiber optic bundle is used to produce a spatially uniform irradiation beam. Visible light filtered through a 380 ± 50 nm bandpass filter passes into the chamber and excites the luminescent sample.

The luminescence emission is collected with a lens, filtered by a 650 ± 20 nm bandpass filter, focused onto a photodetector and imaged onto a cooled CCD camera. The luminescence signal is sampled by a personal computer when the pressure is varied. Thus, the relation between the luminescence intensity and pressure can be obtained at constant temperature. The calibration data can be fitted by the Stern-Volmer equation. The shooting condition is a shutter opening of F22 and exposure time of 150 ms. The total number of pixels is 1008 x 1018.

Figure 4.29 shows luminescence ration images of the upper surface from three angles of attack. These qualitative flow visualizations clearly show a small expansion region near the nose tip and a change in the shock wave location and shape of the separated region behind the shock wave. It should be noted that the model shape in the image also changes with respect to the angle of attack.

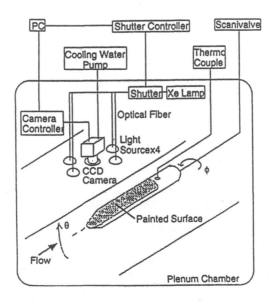

Fig. 4.28 Experimental setup for the Transnoic Wind Tunnel. (Reproduced with permission by Y. Shinbo et al., *20th AIAA Advanced Measurement and Ground Testing Technology Conference*, June

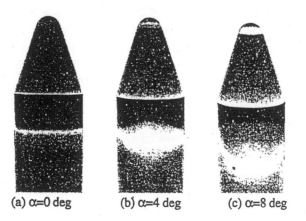

(a) α=0 deg   (b) α=4 deg   (c) α=8 deg

Fig. 4.29   Luminescence ratio image of the H2 rocket model by using the PtOEP system.

References

1. D. A. Skoog, D. M. West and F. J. Holler, *Fundamentals of Analytical Chemistry*, Saunders, Philadelphia, 5th ed., 1988, p. 344.
2. L. C. Clark, US Pat., 2 913 386 (1959).
3. T. M. Freeman and W. R. Seitz, *Anal. Chem.,* 53 (1981) 98.
4. H. D. Hendricks, US Pat., 3 709 663 (1973).
5. X. M. Li and H. Y. Wong in Transient D.O. *Measurement Using a Computerized Membrane Electrode.* Horizons of Biochemical Engineering, ed. Aiba, S., University of Tokyo Press, Tokyo, 1987, p. 213.
6. X. M. Li and H. Y. Wong, US Pat., 4 921 582 (1990).
7. G. J. Mohr and O. S. Wolfbeis, *Anal. Chim. Acta,* 316 (1995) 239.
8. A. V. Vaughan, M. G. Baron and R. Narayanaswamy, *Anal. Comm.,* 33 (1996) 393.
9. H. N. McMurray, P. Douglas, C. Busa and M. S. Garley, *J. Photochem. Photobiol. A: Chem.,* 80 (1994) 283.
10. S. M. Ramasamy and R. J. Hurtubise, *Anal. Chim. Acta,* 152 (1983) 83.
11. M. R. Surgi, *Applied Biosensors,* ed. D. L. Wise, Butterworth, Guildford, 1989, Ch. 9.
12. J. M. Vanderkooi and D. F. Wilson, *Oxygen Transport to Tissue VIII.* I. S. Longmuir (Ed.), Plenum, New York, (1986) 189.
13. J. I. Peterson, R. V. Fitzgerald and D. K. Buckhold, *Anal. Chem.,* 56 (1984) 62.
14. J. L. Gehrich, D. W. Lbers, N. Opitz, D. R. Hansmann, W. W. Miller, J. K. Tusa and M. Yafuso, *IEEE Trans. Biomed. Eng., BME*-33 (1986) 117.
15. G. N. Lewis and M. Kasha, *J. Am. Chem. Soc.,* 66 (1944) 2100.
16. P. F. Lott and R. J. Hurtubise, *J. Chem. Edu.,* 51 (1974) A315.
17. S. Fischkoff and J. M. Vanderkooi, *J. Gen. Physiol.,* 65 (1975) 663.
18. M. Smoluchowski, *Z. Phys. Chem.,* 92 (1917) 129.
19. W. K. Subczynski and J. S. Hyde, *Biophys. J.,* 45 (1984) 743.
20. J. M. Vanderkooi, G. Maniara, T. J. Green and D. F. Wilson, *J. Biol. Chem.,* 262 (1987) 5476.
21. H. Kautsky and A. Hirsch, *Z. Anorg. Allg. Chem.,* 222 (1935) 126.
22. H. Kautsky and G. O. Mler, *Z. Naturforsch,* 2A (1947) 167.
23. H. Kautsky, A. Hirsch and F. Davidsher, *Ber. Dtsch. Chem. Ges.,* 65 (1932) 1762.
24. I. A. Zakharov and T. I. Grishaeva, *Zhur. Prikl. Specktrosk.,* 36 (1982) 980, Engl. Ed. p. 697.
25. I. Bergmann, *Nature,* 218 (1968) 396.
26. O. S. Wolfbeis, H. Offenbacher, H. Kroneis and H. Marsoner, *Mikrochim Acta.,* I (1984) 153.
27. H. W. Kroneis, H. J. Marsoner, *Sens. Actuators,* 4 (1983) 587.
28. D. Lber and N. Opitz, *Sens. Actuators,* 4 (1983) 641.
29. M. E. Cox and D. Dunn, *Appl. Optics,* 24 (1985) 2114.

30. O. S. Wolfbeis, H. E. Posch and H. Kroneis, *Anal. Chem.,* 57 (1985) 2556.

31. I. S. Longmuir and J. A. Knopp, *J. Appl. Physiol.,* 41 (1976) 598.

32. M. H. Mitnick and F. F. Jis, *J. Appl. Physiol.,* 41 (1976) 593.

33. M. Kautsky, *Trans. Faraday Soc.,* 35 (1939) 216.

34. X. M. Li and K. Y. Wong, *Anal. Chim. Acta,* 262 (1992) 27.

35. Y. M. Liu, R. Pereiro-Garcia, M. J. Valencia-Gonzalez, M. E. Diaz-Garcia and A. Sanz-Medel, *Anal. Chem.,* 66 (1994) 836.

36. J. M. Charlesworth, *Sens. Actuators B,* 22 (1994) 1.

37. N. Velasco-Garcia, R. Pereiro-Garcia and M. Diaz-Garcia, *Spectrochim. Acta,* 51A (1995) 895.

38. D. M. Papkowsky, G. V. Ponomarev, W. Trettnak and P. O'Leary, *Anal. Chem.,* 67 (1995) 4112.

39. S.-K. Lee and I. Okura, *Spectro. Chim. Acta A.*54 (1998) 91.

40. O. S. Wolfbeis, *Fiber Optic Chemical Sensors and Biosensors,* Vol. 2, CRC Pressure, Boca Raton FL, pp. 19-53 (1991).

41. D. B. Papkovsky, G. V. Ponomarev, W. Trettnak and P. O'Leary, *Anal. Chem.,* 67 (1995) 4112.

42. W. Xu, K. A. Kneas, J. N. Demas and B. A. DeGraff, *Anal. Chem.,* 68 (1996) 2605.

43. J. Samuel, A. Strinkovsk, S. Shalom, K. Lieberman, M. Ottolenghi, D. Avnir and A. Lewis, *Mater. Lett.,* 21 (1994) 431.

44. D. Avinir, D. Levy and R. Reisfeld, *J. Phys. Chem.,* 88 (1984) 5956.

45. R. Zusman, C. Rottman, M. Ottolenghi and D. Avnir, *J. Non-Cryst. Solids,* 122 (1990) 107.

46. G. E. Badini, K. T. V. Grattan and A. C. C. Tseung, *Analyst,* 120 (1995) 1025.

47. I. Kuselman and O. Lev, *Talanta,* 40 (1993) 749.

48. L. Yang and S. S. Saavedra, *Anal. Chem.,* 67 (1995) 1307.

49. K. T. V. Grattan, G. E. Badini, A. W. Palmer and A. C. C. Tseung, *Sens. Actuators A,* 25-27 (1991) 483.

50. U. Narang, P. N. Prasad, F. V. Bright, K. Ramanathan, N. D. Kumar, B. D. Malhotra, M. N. Kamalasanan and S. Chandra, *Anal. Chem.,* 66 (1994) 3139.

51. Better Ceramics Through Chemistry II , ed. C. J. Brinker, D. E. Clark and D. R. Ulrich, *MRS Symposium,* Vol. 73, Elsevier, New York, 1986.

52. C. J. Brinker and G. W. Scherer, *The Physics and Chemistry of Sol-Gel Processing,* Academic Press, New York, 1990.

53. S.-K. Lee and I. Okura, *Analyst,* 122 (1997) 81.

54. Z. Grauer, D. Avinir and S. Yariv, *Can. J. Chem.,* 62 (1984) 1889.

55. D. Avinir, V. R. Kaufman and R. Reisfeld, *J. Non-Cryst. Solids,* 74 (1985) 395.

56. A. K. McEvoy, C. M. McDonagh and B. D. MacCraith, *Analyst,* 121 (1996) 785.

57. D. B. Papkovsky, *Sens. Actuators B,* 11 (1993) 293.

58. H.W. Kroneis and H.J. Marsoner, *Sens. Actuators,* 4 (1983) 587.

59. J.I. Peterson, R.V. Fitzgerald and D.K. Buckhold, *Anal. Chem.,* 56 (1984) 62.

60. S.-K. Lee and I. Okura, *Anal. Chim. Acta,* 342 (1997) 181.

61. Y. Amao and I. Okura, *Analyst,* 125 (2000) 1601.

62. S.-K. Lee and I. Okura, *Anal. Commun.,* 34 (1997) 185.

63. S.-K. Lee and I. Okura, *Anal. Sci.,* 13 (1997) 181.

64. E.R. Carraway, J.N. Demas, B.A. DeGraff, and J.R. Bacon, *Anal. Chem.,* 63 (1991) 332.

65. X.-M. Li, F.-C. Ruan and K.-Y. Wong, *Analyst,* 118 (1993) 289.

66. M. Gouterman, The Porphyrins Vol.3. Academic Press, New York, 1978.

67. K. Arishima, H. Hiratsuka, A. Tate, and T. Okada, *Appl. Phys. Lett.,* 40(1982) 279.

68. T.A. Temofonte and K.F. Schoch, *J. Appl. Phys.,* 65 (1989) 1350.

69. R.H. Moser and A.L. Thomas *The Phthalocyanines;* CRC Press: Boca Raton, FL, 1983; Vol I and II.

70. J.H. Brannon and D. Magde, *J. Am. Chem. Soc.,* 102 (1980) 62.

71. H. Ohtani, T. Kobayashi, T. Tanno, A. Yamada, D. Wohrle and T. Ohno, *Photochem. Photobiol.,* 44 (1986)125.

72. T. Ohno, S. Kato, A. Yamada and T. Tanno, *J. Phys. Chem.,* 87 (1983) 775.

73. S. Draxler and M.E. Lippitsch, *Anal. Chem.,* 68 (1996) 753.

74. M.E. Lippitsch, J. Pusterhofer, M.J.P. Leiner, O.S. Wolfbeis, *Anal. Chim. Acta,* 205 (1988) 1.

75. J. McVie, R. S. Sinclair and T. G. Truscott, *J. Chem. Soc. Faraday Trans.* 2 (1978) 1870.

76. P. Jacques and A. M. Braun, *Helv. Chim. Acta,* 64 (1981) 1800.

77. B. Amand and R. Bensasson, *Chem. Phys. Lett.,* 34 (1975) 44.

78. T. Furuto, S-K. Lee, K. Asai, and I. Okura, *Chem. Lett.,* 61 (1998).

79. I. A. Silver, *Polarographic Techniques of Oxygen Measurements in Oxygen: An In-Depth Study of Its Pathophysiology*, S. F. Gottlieb, I. S. Longmuir and J. R. Totter, eds., Undersea Med. Soc. Pub No. 62 (ws) 3-1-84 p. 215-238.

80. J. M. Vanderkooi, G. Maniara, T. J. Green and D. F. Wilson, *J. Biol. Chem.,* 262 (1987) 5476.

81. W. L. Rumsey, J. M. Vanderkooi and D. F. Wilson, *Science,* 241 (1988) 1649.

82. D. F. Wilson and G. J. Cerniglia, *Cancer Res.,* 52 (1992) 3988.

83. H.C. Gerritsen, R. Sanders, A. Draaijer, C. Ince and Y.K. Levine, *J. Fluoresc.,* 7 (1997) 11.

84. K. Kalyanasundaram and M. Neumann-Spallart, *J. Phys. Chem.,* 86 (1982) 5163.

85. L.-W. Lo, C. J. Koch and D. F. Wilson, *Anal. Biochem.,* 236 (1996) 153.

86. M.J. Morris, J.F. Donovan, J.T. Kegelman, S.D. Schwab, R.L. Levy and R.C. Crites, *30th Aerospace Sciences Meeting and Exhibit,* Reno, NV, USA, Jan. 6-9, 1992.

87. J. Kavandi, J. Callis, M. Gouterman, G. Khalil, D. Wright, E. Green, D. Burus and B. McLachlan, *Res. Sci. Instrum.,* 61 (1990) 3341.

88. I. Troyanovsky, N. Sadovskii, M. Kuzmin, V. Mosharov, A. Orlov, V. Radchenko and S. Phonov, *Sens. Actuators B,* 11 (1993) 201.

89. O. S. Wolfbeis, *Fiber Optic Chemical Sensors and Biosensors,* Vol. 2, CRC Pressure, Boca Raton FL, pp. 19-53 (1991).

90. K. Asai, H. Kanda and Y. Iijima, *J. Visual. Soc. Jpn.,* 15 (1995) 59.
91. S-K. Lee and I. Okura, *Anal. Sci.,* 13 (1997) 535.
92. W. Xu, K.A. Kneas, J.N. Demas and B.A. DeGraff, *Anal. Chem.,* 68 (1996) 2605.
93. Y. Shinbo, K. Asai, H. Kanda, Y. Iijima, N. Komatsu, S. Kita and M. Ishiguro, *20th AIAA Advanced Measurement and Ground Testing Technology Conference*, June 15-18, 1998.

Chapter 5   Sensitizers for Photodynamic Therapy

5.1   Introduction to Photodynamic Therapy (PDT)

Photodynamic therapy (PDT) is a phototherapeutic method for destroying solid tumors by irradiation after the injection of a tumor localized photosensitizer. This method requires a photosensitizer that can accumulate selectively in tumors and localized irradiation. PDT is superior to chemotherapy and radiotherapy. The photosensitizers employed must exhibit good phototoxicity and low dark toxicity. This method of cancer treatment seems to be very promising, because dangerous side effects in other tissues as known in chemotherapy and radiotherapy can be avoided if the photosensitizer is selectively accumulated in tumor tissues. As a result of recent developments in laser techniques and improvements in optical fibers, the application is now expanding.

The fluorescence of the photosensitizer incorporated in tumor tissues may be detected after irradiation and can be used as a diagnostic means to probe the localization of the photosensitizer.

The photosensitizer preparation used almost exclusively in PDT consists of a mixture of hematoporphyrin derivatives, HpD. Cancer therapy using hematoporphyrin was carried out by Diamond et al. in 1972[1], and PDT with HpD was started by Dougherty et al. in 1978.[2] The most active compounds in this mixture are believed to be dihematoporphyrin ethers and/or esters. A commercial preparation enriched in the latter, Photofrin, has been made available for research and clinical trials.[3]

5.1.1 Method of PDT

The treatment of the patient is carried out in the following steps (the sequence of events in PDT is shown in Scheme 5-1).[4] First, the solution of a suitable photosensitizer is injected systematically, then time is allowed for tumor accumulation of the photodrug (photosensitizer). After 24 - 72 h the concentration of the photosensitizer in the tumor is often higher than in normal tissue. Irradiation with an appropriate dose of visible light results in the activation of the photodrug and subsequent

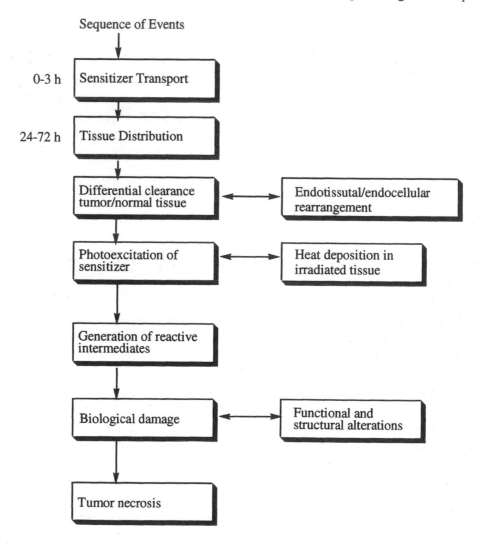

Scheme 5-1 Sequence of events in photodynamic therapy. (Reproduced with permission by G.Jori, *Photochem. Photobiol.*, **52** (1990) 439, American Society for Photobiology)

photoreactions with parts of the tumor tissue within the first few hours after photodynamic treatment. Profound cellular effects occur within 12 to 18 h, causing a number of processes. After 2 - 3 weeks necrosis of the tumor tissue takes place.

5.1.2 Reaction mechanisms of PDT

The principal cause of photodamage in PDT involves the following processes.[5] By irradiation the photosensitizer is photoexcited to the triplet state via an intersystem crossing from the singlet state to the triplet state, as shown in Fig. 5.1. To be effective the photosensitizer must possess a long triplet state lifetime. The key chemical species inflicting cellular damage are formed by the following two reactions.

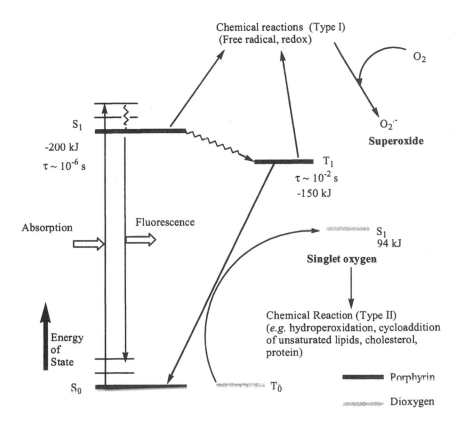

Fig. 5.1 Generation of excited porphyrin states and reactive dioxygen species. State energies are represented by thick lines; ▬▬porphyrin sensitizer, ▬▬ dioxygen. Reactive dioxygen intermediates are in bold type. (Reproduced with permission by A. W. Girotti, *Photochem. Photobiol.*, **38** (1983) 745, American Society for Photobiology)

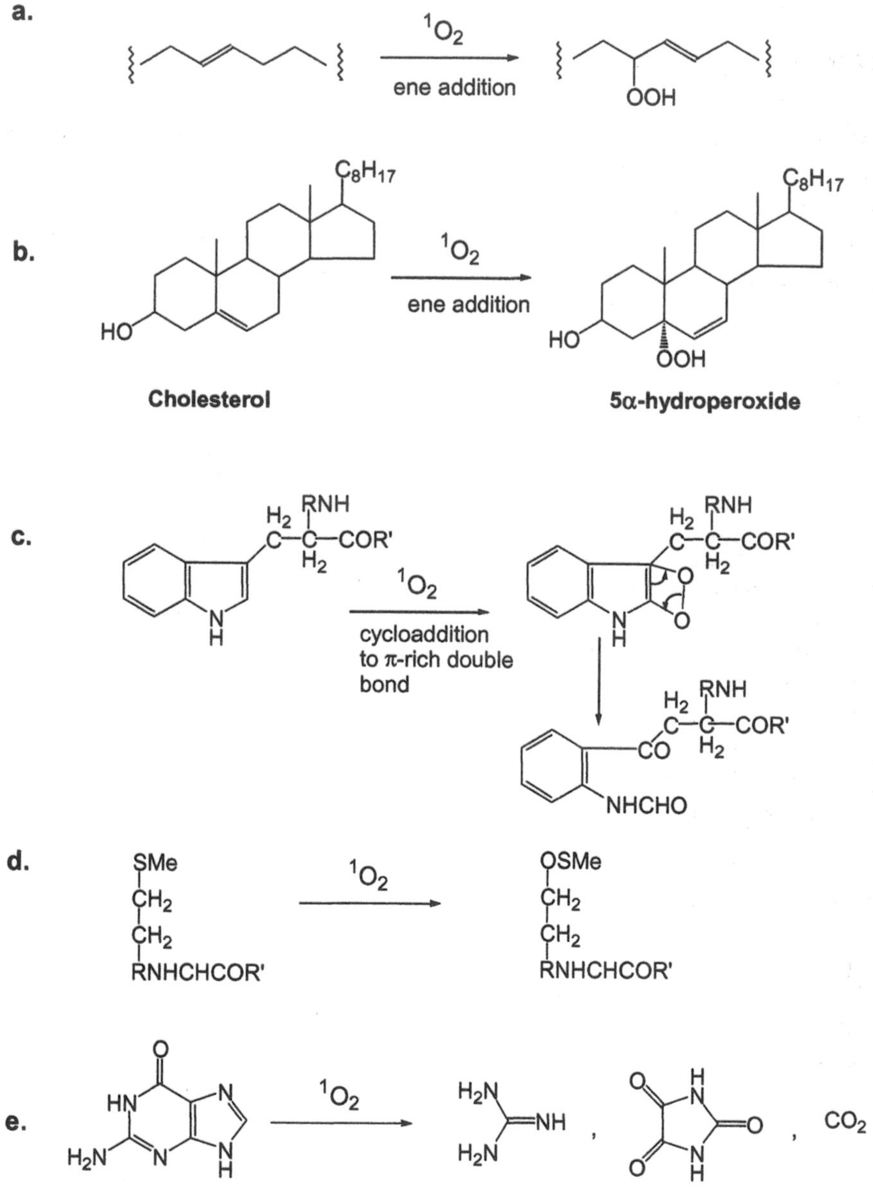

Fig. 5.2   Reactions of singlet oxygen with some biomolecules. a; unsaturated lipid, b; cholesterol, c; tryptophan, d; methionine, e; guanine. (Reproduced with permission by A. W. Girotti, *Photochem. Photobiol.*, **38** (1983) 745, American Society for Photobiology)

Type I reaction: This reaction is the photoinduced electron transfer (redox reactions) either as reductive quenching by biomolecules or oxidative quenching by oxygen to form radical species including superoxide.

Type II reaction: This reaction is the photoinduced energy transfer to form singlet oxygen.

It is generally agreed that singlet oxygen is the key agent of cellular damage. In the case of Cu-benzochlorine, however, the reaction proceeds via a Type I reaction. Singlet oxygen is a powerful oxidant that reacts with a variety of biomolecules and assemblies, as shown in Fig. 5.2. Although the superoxide is less reactive, a very reactive hydroxy radical is formed by the Harber-Weiss reaction.

5.1.3 Photosensitizers

The photosensitizer used in PDT was almost exclusively HpD, which was synthesized by Lipson et al. in 1961.[6] HpD, a mixture of porphyrin derivatives (the structure is shown in Fig. 5.3), is

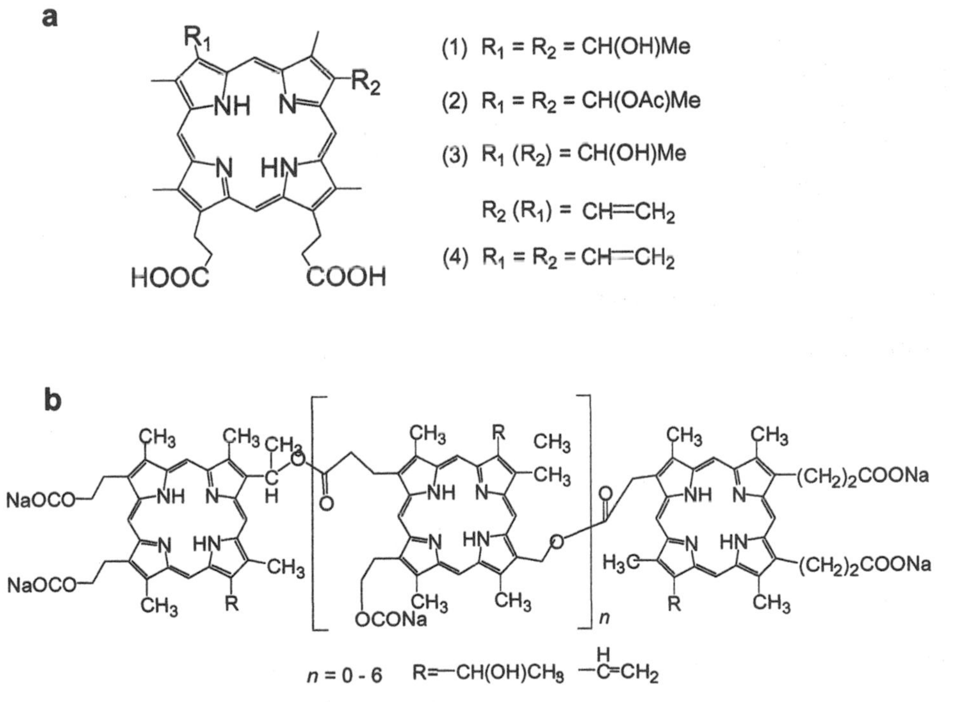

Fig. 5.3 Structure of HpD. **a.** Porphyrin monomers containing HpD (1) hematoporphyrin, (2) hematoporphyrin diacetate, (3) 3(8)-hydroxylethyl-8(3)-vinyldeuteroporphyrin, (4) protoporphyrin. **b.** Higher molecular weight material in HpD (porphyrin dimers and oligomers with ether, ester and carbon-carbon linkages).

synthesized by the treatment of hematoporphyrin with H_2SO_4-CH_3COOH. The main component in this mixture is hematoporphyirn diacetate. For injection the mixture is neutralized. By neutralization hematoporphyrin, vinyldeuteroporphyrin and protoporphyrin are formed. Besides these compounds some covalent oligomers such as dihematoporphyrin ethers and/or esters are formed. Since the oligomeric components show the highest affinity to tumor tissues, a partially purified version of HpD called Photofrin of similar compounds is employed.

Red light is used for PDT because of the optimal penetration through tissues and compatibility with laser technologies. However, HpD and Photofrin absorb the light only weakly above 600 nm. Since these compounds exhibit some disadvantages new photosensitizers which strongly absorb light in longer wavelengths are being developed. Several classes of dyes such as chlorin, purpurin, benzoporphyrin, phthalocyanine and naphthalocyanine, with absorption maxima well above 650 nm, have been advanced as second-generation photosensitizers, as shown in Figs. 5.4 and 5.5.[7-9] The lifetime of the photoexcited triplet state and the quantum yield of singlet oxygen formation of the second-generation photosensitizers are superior to first-generation photosensitizers such as HpD and Photofrin.

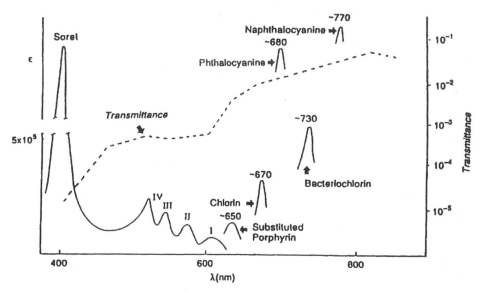

Fig. 5.4   Photosensitizer absorbance in relation to tissue transmittance. The absorption spectra are schematic: only Band I is shown, except for the porphyrin absorprion spectrum on the left. The transmittance curve refers to a sample of human scrotal sac, 7 mm thick. The broad feature at 500 - 600 nm is attributed to absortion by hemoglobin. (Reproduced with permission by S. Wan, J. A. Parrish, R. R. Anderson, M. Madden, *Photochem. Photobiol.*, **34** (1981) 679, American Society for Photobiology)

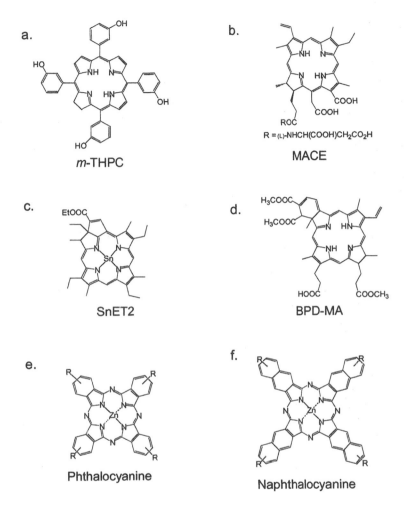

Fig. 5.5 Structures of the second generation photosensitizers.

5.2   Porphyrins for PDT

As noted above, the photosensitizers used in PDT have been almost exclusively HpD and Photofrin in the clinic. Although the new second-generation photosensitizers are not always porphyrins, unusual porphyrins for PDT are described in this section.

5.2.1 Porphyrin oligomer

HpD and Photofrin are known to be mixtures of monomeric and aggregated porphyrins. When a model compound, porphyrin oligomer (the structure is shown in Fig. 5.6) is added to the growth medium during HeLa cell cultivation, porphyrin oligomer is incorporated into HeLa cells.[10] Although porphyrin oligomer does not fluoresce in aqueous solution, the porphyrin incorporated in HeLa cells shows strong fluorescence. The porphyrin oligomer aggregates or is folded in aqueous solution. In HeLa cells aggregated porphyrin dissociates. The porphyrin concentration in HeLa cell increases almost linearly with contact time and the fluorescence intensity strongly

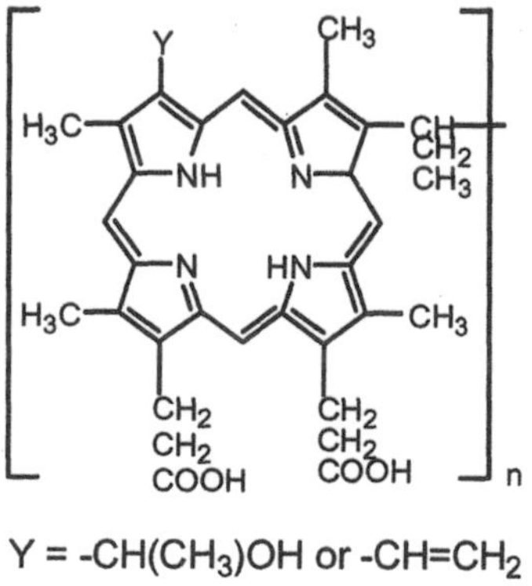

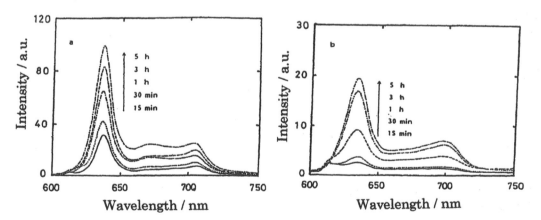

$Y = -CH(CH_3)OH$ or $-CH=CH_2$

Fig. 5.6 Proposed structure of porphyrin oligomer.

Fig. 5.7 Time dependence of fluorescence spectrum from porphyrin oligomer (a) or Photofrin II (b) incorporated in HeLa cells. See text for reaction conditions.

depends on the incubation time, as shown in Fig. 5.7. Upon irradiation of HeLa cells (by argon laser) containing the porphyrin oligomer, membrane damage is observed. By microscopic observation the cells change gradually, as shown in Fig. 5.8. Just after irradiation a foamy appearance in HeLa cells is observed, the shape changes to become globular with time, and the ciliary process on the cell surface disappears. In the presence of trypan blue, HeLa cells are gradually colored with time, showing cell membrane damage.

In *in vivo* experiments when porphyrin oligomer is injected intraperitoneally to MH-134 ascites cell-implanted C3H/He mice, the fluorescence intensity of the porphyrin oligomer incorporated in the ascites cell increases with time, and fast development of intense fluorescence is observed. Comparing porphyrin oligomer and Photofrin, faster development of more intense fluorescence is observed for porphyrin oligomer, as shown in Fig. 5.9.

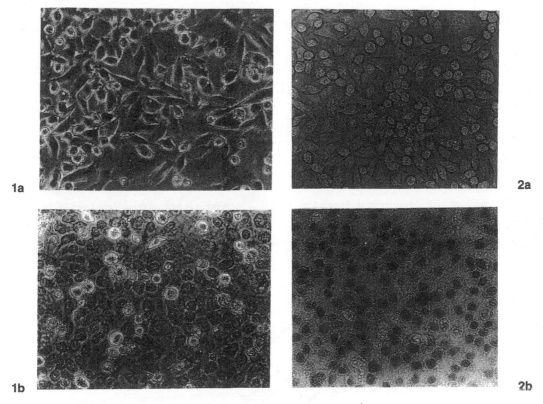

Fig 5.8   Microscopic pictures of HeLa cells stained with trypan blue obtained by phase contrast microscope. 1a: HeLa cells are contacted with porphyrin oligomer, 1b: 1a + irradiation 2a: HeLa cells are contacted with Photofrin II, 2b: 2a + irradiation. (Reproduced with permission from *Lasers in the Life Sciences*, **4** (1991) 87)

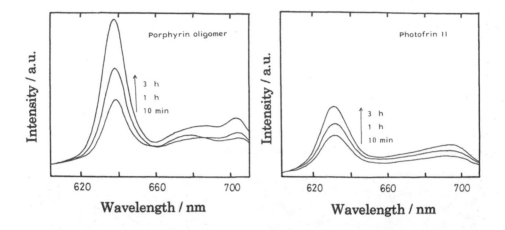

Fig. 5.9   Fluorescence spectra of Photofrin II in 5 vol% Triton X-100 (a) and in aqueous solution
(b), and porphyrin oligomer in 5 vol% Triton X-100 (c) and in aqueous solution (d).
(Reproduced with permission from *Lasers in the Life Sciences*, **4** (1991) 87)

5.2.2 Zincphyrin

The structure of zincphyrin is shown in Fig. 5.10.[11] Zincphyrin is primarily isolated as a histamine-
release inhibitor from mast cells. In extensive pharmacological studies, Zincphyrin has been
found to have potent photosensitizer activity. Zincphyrin is obtained from the supernatant of the

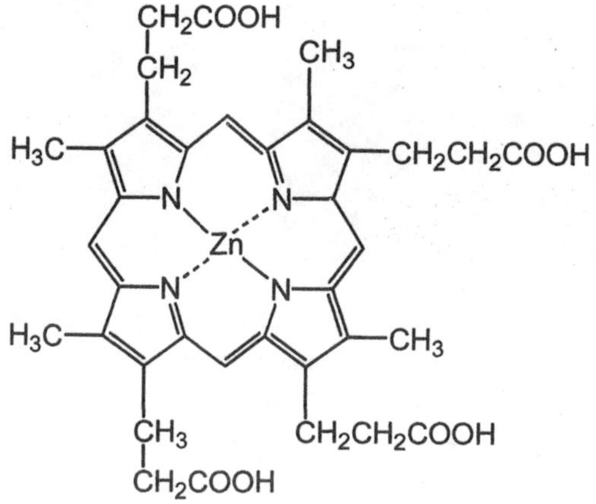

Fig. 5.10 The structure of zincphyrin.

culture broth of *Streptomyces* AC8007 (FERM BP-10537).

Zincphyrin is compared to HpD, and PDT is carried out with sarcoma 180 (S-180)-bearing mice. S-180 cells are transplanted into the shaven back of male ICR mice by subcutaneous injection of approximately 5×10^6 cells per mouse. The tumor grows about 130 mg within two days after implantation. Zincphyrin and HpD (50 mg/kg ip) dissolved in physiological saline containing 0.1 M Tris buffer (pH 7.4) are given to mice in groups of five. Ten minutes later, the tumors are irradiated with 75.48 mW cm^{-2} white light derived from halogen lamp for 10 min. Zincphyrin is found to be effective as a photosensitizer to control tumor growth and is less toxic than HpD in mortality of mice (Fig. 5.11). Zincphyrin is considered to be a photosensitizer having high purity and low toxicity.

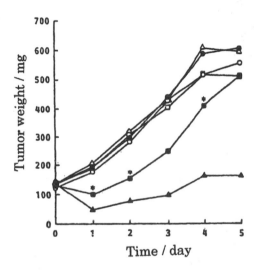

Fig. 5.11 Effect of zincphyrin and HpD on growth of sarcoma 180. ○ Control, ● irradiation, zincphyrin, H zincphyrin+irradiation, □ HpD, ■ HpD+irradiation. Each point represents the mean of five mice. *One mouse died. (Reproduced with permission by M. Toriya et al., *J. Antibiotics*, **46** (1993) 196)

5.3   Phthalocyanines for PDT

Photodynamic therapy utilizing porphyrin as a photosensitizer is being widely investigated for the treatment of a variety of tumors. In this therapy, it is important to accumulate porphyrin in tumors selectively. Although HpD and Photofrin have been used exclusively in photodynamic therapy, the development of more effective photosensitizers is required. When light with an absorption band of longer wavelength is used, the tissue penetration is deeper. Photosensitizers with relatively long absorption wavelength are suitable for effective photodynamic action. Metallophthalocyanines show absorption maximum around 650-700 nm, and Ga-, Al-, and Zn-phthalocyanines have been used as photosensitizers. Various types of phthalocyanines are employed in PDT. Their structures are shown in Fig. 5.12. Water-soluble zinc tetrasulfophthalocyanine, for example, shows its maximum absorption band around 670 nm. Its wavelength has relatively long penetration depth and thus seems to be a more suitable photosensitizer than porphyrin for PDT.

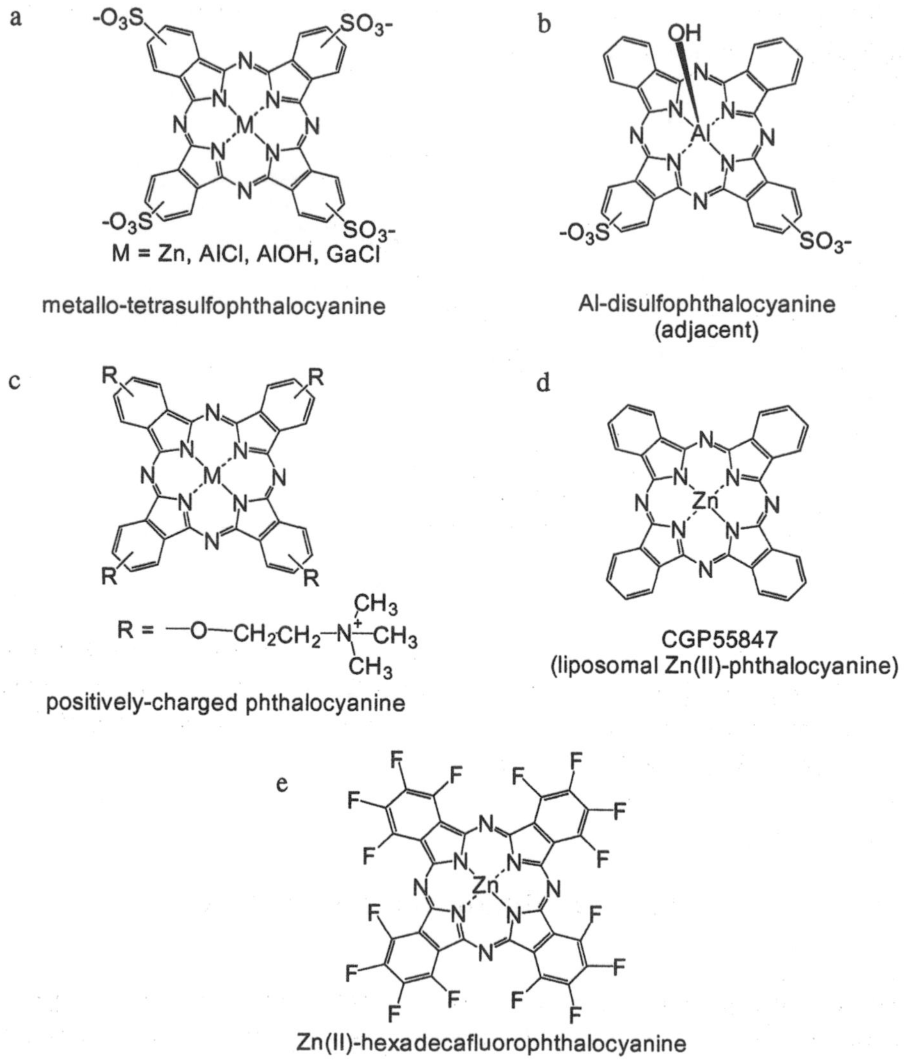

a

M = Zn, AlCl, AlOH, GaCl

metallo-tetrasulfophthalocyanine

b

Al-disulfophthalocyanine
(adjacent)

c

$R = -O-CH_2CH_2-N^+(CH_3)_3$

positively-charged phthalocyanine

d

CGP55847
(liposomal Zn(II)-phthalocyanine)

e

Zn(II)-hexadecafluorophthalocyanine

Fig. 5.12 Structures of various phthalocyanines used in photodynamic therapy.

5.3.1 Zinc-sulfophthalocyanine

Among metallophthalocyanines zinc-sulphophthalocyanine is one of the suitable photodrugs for PDT for the following reasons. 1) Zinc phthalocyanine is water-soluble, which is necessary for injection. 2) Photoexcited triplet state lifetime is long enough (*ca.* 100 μs) to form key species such as singlet oxygen to inflict cellular damage. 3) The maximum absorption band is in the long wavelength region (*ca.* 670 nm).

Zinc phthalocyanines with different degrees of sulfonation have been synthesized (ZnTSPc and ZnSPc$_{mix}$, as shown in Fig. 5.13) and are compared for tumor uptake, affinity to protein and

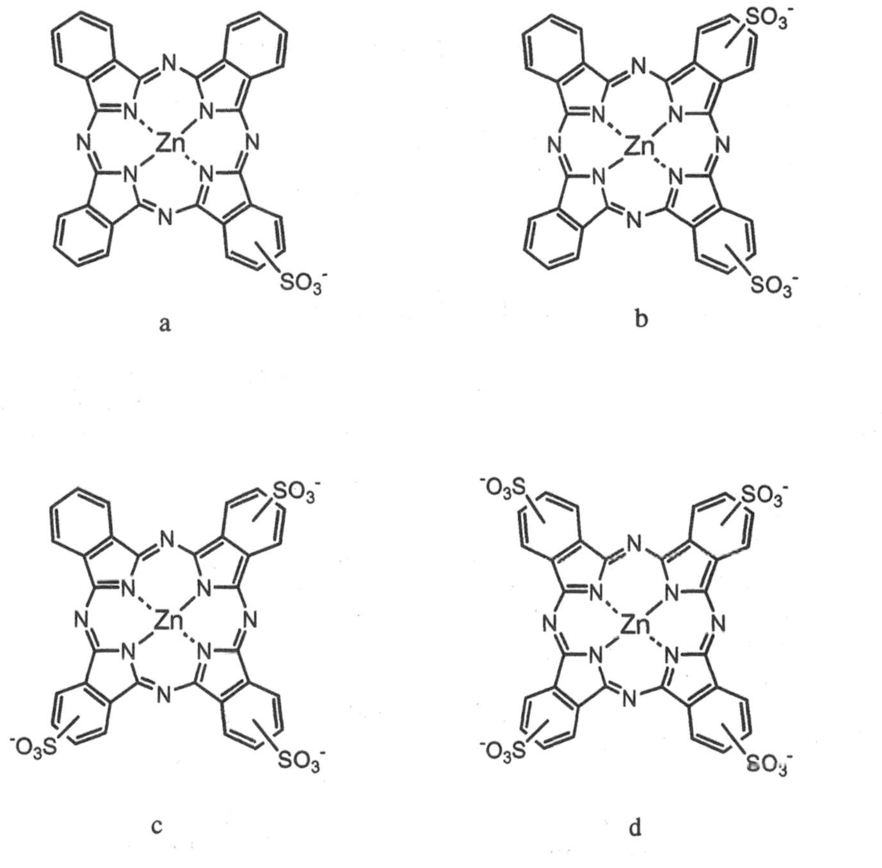

Fig. 5.13 Structures of sulfonated zinc phthalocyanines. (a) mono-sulfonated, (b) di-sulfonated, (c)tri-sulfonated, (d) tetra-sulfonated (ZnTSPc). ZnSPc$_{mix}$ contains (a), (b), (c) and (d).

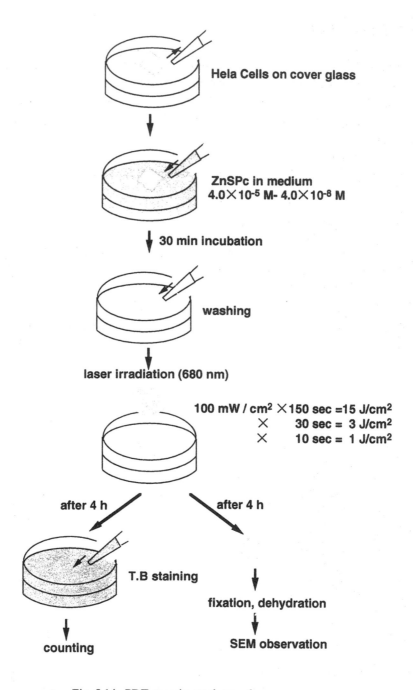

Fig. 5.14 PDT experimental procedure.

phototoxicity. ZnTSPc and ZnSPc$_{mix}$ are synthesized by refluxing a mixture of the following compounds: monosodium salt of 4-sulfophthalic acid, ammonium chloride, urea, ammonium molybdate, and zinc(II) sulfate 7-hydrate. The crude product is added to 600 ml of 1 N hydrochloric acid saturated with sodium chloride. The slurry is heated at 80°C until the ammonia evolution stops. After cooling, the solution is neutralized with NaOH solution, filtrated and the precipitate washed with methanol. The product is then purified by ODS silica gel to remove sodium chloride.

The PDT experimental procedure is shown in Fig. 5.14. Phthalocyanines are contacted with the HeLa cells cultivated on the cover glass. After 30 min the cells are washed three times. The cellular uptake is measured by the excitation spectrum at 680 nm. When HeLa cells are used the cellular uptake of ZnSPc$_{mix}$ is better than that of ZnTSPc, as shown in Fig. 5.15.

After irradiation of HeLa cells by DCM dye laser, the cellular damage is measured by trypan blue method and the results are shown in Table 5.1. In the case of ZnTSPc, phototoxicity decreases with decreasing irradiation energy at a concentration of 4.0×10^{-5} mol dm^{-3}. Under 4.0×10^{-6} mol dm^{-3}, no phototoxicity is observed. In comparison, the phototoxicity by ZnSPc$_{mix}$ is observed even at a dye concentration of 4.0×10^{-7} mol dm^{-3}. Although the dye concentration and irradiation energy are low, HeLa cells are damaged by irradiation with ZnSPc$_{mix}$. Since ZnSPc$_{mix}$

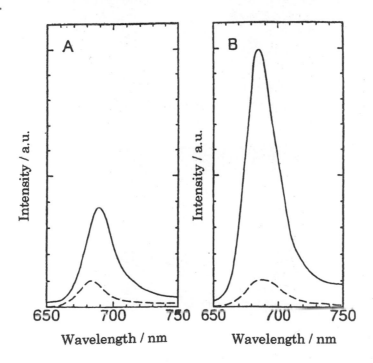

Fig. 5.15 Fluorescence emission spectra of ZnTSPc (A:——) and ZnSPc$_{mix}$ (B: • • •). After addition of 5% FBS, fluorescence intensities of ZnTSPc (A:——) and ZnSPc$_{mix}$ (B:• • •) were changed. Excitation wavelength was 620 nm.

202

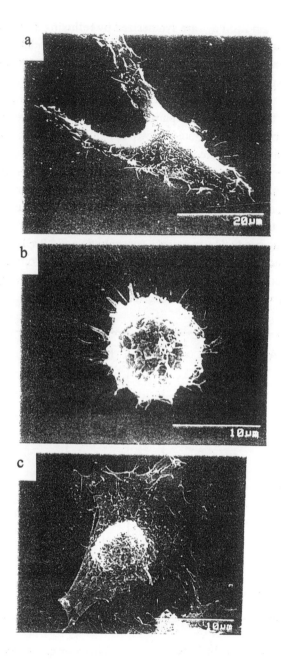

Fig. 5.16 SEM observation of HeLa cell. a: HeLa cell before irradiation. b: HeLa cell was contacted with ZnTSPc and irradiated. c: HeLa cell contacted with $ZnTSPc_{mix}$ and irradiated. (Reproduced with permission from *Lasers in the Life Sciences*, **6** (1994) 173)

Table 5.1   Effect of PDT

	Irradiation energy / J· cm^{-2}	conc./dm^{-3}			
		4.0×10^{-5}	4.0×10^{-6}	4.0×10^{-7}	4.0×10^{-8}
ZnTSPs	15	100	0	0	-
	3	91	0	0	-
	1	30	0	0	-
ZnDSPc	15	100	100	100	0
	3	100	100	100	0
	1	100	100	90	0

(Reproduced with permission from *Lasers in the Life Sciences*, **6** (1994) 173)

has hydrophobic parts in the molecule, they can more easily come into the cell membrane than ZnTSPc.

Figure 5.16 shows the morphological change of a HeLa cell by scanning electron microscope. Fig. 5.16a shows a regular HeLa cell. Fig. 5.16b shows the morphological change in HeLa cell after PDT with ZnTSPc. This change progressed with time. At 4 h after laser irradiation, the HeLa cell is round, as shown in the figure. Fig. 5.16c shows the morphological change in the cell immediately after PDT with ZnSPc$_{mix}$. Compared with Fig. 5.16a and b, the nuclear material stands out and the outline of cell is consistent due to cytolysis. The cell membrane is strongly damaged by the irradiation, only the membrane is disrupted by the phototoxic species formed by the photosensitization of phthalocyanines adsorbed on the cell membrane.

5.3.2 New phthalocyanines for PDT

HpD and Photofrin exhibit low absorption in the red region of the spectrum. PDT could be rendered more efficient through the use of photosensitizers which absorb more strongly towards the red end of the visible spectrum, because the light above 680 nm allows for deeper penetration into biological tissues. A variety of porphyrins and porphyrin-like compounds have been examined in the course of this search.[12-24] Phthalocyanines exhibit advantageous photophysical properties for PDT including photostability, a long lifetime of the photoexcited triplet state and a high molar absorption in the red region of the visible spectrum.

Phthalocyanines exhibit absorption bands near 670 nm. However, it is preferable that photosensitizers exhibit longer absorption bands for the reason stated above. The water-soluble fluorinated zinc phthalocyanines (Fig. 5.17) exhibit absorption maxima higher than other zinc phthalocyanines. In this section, the photochemical properties of these phthalocyanines and their photodynamic efficiency for HeLa cells are discussed.

A. Photochemical Properties of the Phthalocyanines
Uv-vis spectra of the fluorinated phthalocyanines and ZnTSPc are shown in Fig. 5.18. These zinc

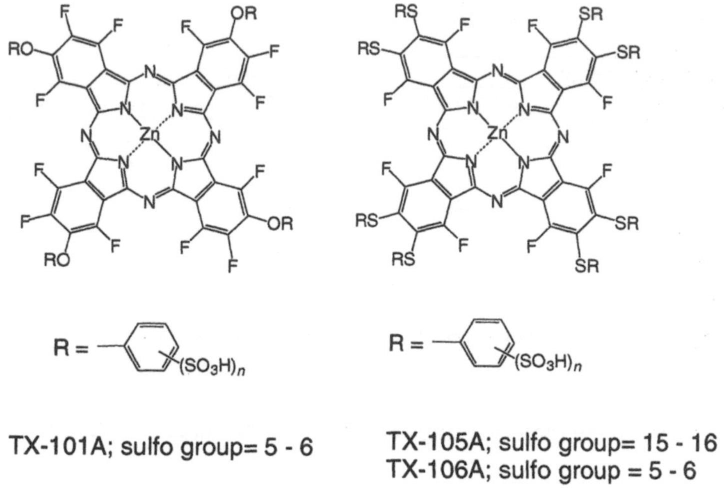

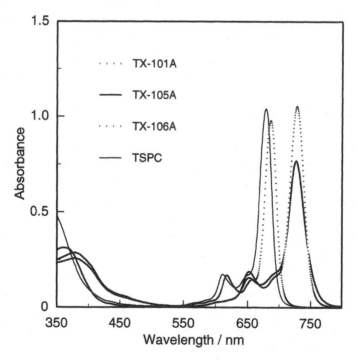

R = —⟨benzene ring⟩(SO₃H)ₙ   R = —⟨benzene ring⟩(SO₃H)ₙ

TX-101A; sulfo group= 5 - 6   TX-105A; sulfo group= 15 - 16
                                      TX-106A; sulfo group = 5 - 6

Fig. 5.17   Structures of fluorinated water-soluble zinc phthalocyanines.

Fig. 5.18   Absorption spectra of fluorinated water-soluble zinc phthalocyanines and TSPC in 1%
Tween 20. (Reproduced with permission from *J. Porphyrins Phthalocyanines*, **2** (1998)
219, John Wiley & Sons Limited)

phthalocyanines easily aggregated in water and dissociated in detergents and some organic solvents. The absorption maxima of TX-101A, TX-105A and TX-106A were 687 nm, 726 nm and 726 nm, respectively, in 1% Tween 20 solution. These maxima are higher than that of ZnTSPc. TX-105A and TX-106A, especially, exhibit absorption maxima at very long wavelengths.

In PDT, the cytotoxic species are formed by energy transfer from the photoexcited triplet state of the sensitizer. It is generally accepted that singlet oxygen is the main cytotoxic agent in PDT. Therefore, lifetimes of the triplet state and quantum yield of singlet oxygen were significantly related to photodynamic efficiencies. The lifetimes of the triplet state and the quantum yield of singlet oxygen generation of the phthalocyanines are shown in Table 5.2. The triplet states of the fluorinated phthalocyanines exhibit long lifetimes. The lifetimes of the triplet state of the fluorinated phthalocyanines are slightly longer than that of ZnTSPc. The quantum yields of singlet oxygen generation of 0.30-0.42 are determined for fluorinated phthalocyanines. The quantum yields of singlet oxygen generation of fluorinated phthalocyanines are of the same order of magnitude as ZnTSPc.

Table 5.2 Photochemical properties and partition coefficients of the fluorinated water-soluble phthalocyanines and ZnTSPC

	Triplet lifetime/ns*	Quantum yield of singlet oxygen*
TX-101A	320	0.42
TX-105A	280	0.34
TX-106A	315	0.30
ZnTSPC	245	0.34

* The lifetimes of the triplet state and the quantum yield of singlet oxygen generation are measured in dimethylsulfoxide. (Reproduced with permission from *J. Porphyrins Phthalocyanines*, **2** (1998) 219, John Wiley & Sons Limited)

B. Cell Uptake

The hydrophilicity of the photosensitizers is significantly related to their photodynamic activities, and amphiphilic photosensitizers exhibit good photodynamic activities. All of the fluorinated phthalocyanines exhibit higher hydrophobicity than ZnTSPc. Fluorinated phthalocyanines are amphiphilic compounds with the most hydrophobic compound being TX-101A.

Figure 5.19 shows the time dependence of the intracellular phthalocyanine concentrations in HeLa cells. All phthalocyanines show rapid cell uptake during the first 60 min, followed by slow uptake during the next 180 min. After 240 min, the highest cellular concentration is observed with TX-106A. The uptake of the anionic photosensitizers is related to the change/mass ratio of the photosensitizers. The uptake of fluorinated phthalocyanines is followed by these charge/mass ratios TX-106A, 2.8; TX-101A, 3.3; TX-105A, 5.4. However, the second highest cellular concentration is observed with ZnTSPc despite the fact that ZnTSPc is the most hydrophilic compound.

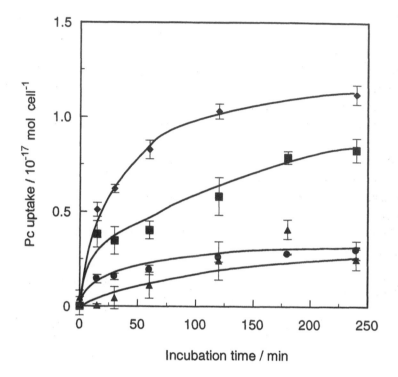

Fig. 5.19 Cellular uptake of phthalocyanines for various time durations. Cells were incubated with
3 x 10⁻⁶ M of TX-101A (●), TX 105A (▲), TX-106A (♦) or ZnTSPC (■). Bars represent
SD of the mean of three experiments. (Reproduced with permission from *J. Porphyrins
Phthalocyanines*, **2** (1998) 219, John Wiley & Sons Limited)

C. Photocytotoxicity

The effect of phthalocyanine concentration on cell survival after irradiation is shown in Fig. 5.20.
Every fluorinated phthalocyanine has greater photodynamic efficiency than ZnTSPc. TX-105A
and TX-106A indicate good photodynamic efficiency (TX-105A, $LD_{90} = 2.1 \times 10^{-6}$ M; TX-106A,
$LD_{90} = 1.6 \times 10^{-6}$ mol dm⁻³).

The difference between TX-105A and TX-106A is only the degree of sulfonation.
However, TX-105A is photodynamically more efficient than TX-106A. The monomeric state of
the molecules is fundamentally important for photoinduced processes since, in the case of associates,
the decay of the triplet state under irradiation occurs as a bimolecular process. Therefore, the
difference in the photodynamic efficiencies between TX-105A and TX-106A may be caused by
the difference in the degree of dissociation of these phthalocyanines in the cell.

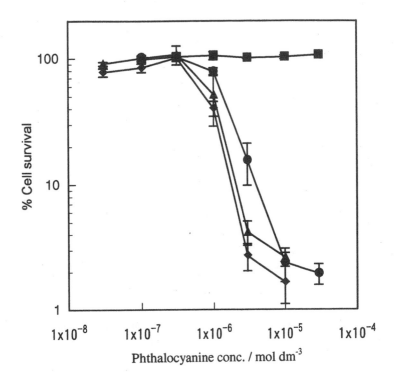

Fig. 5.20 Photocytotoxicities of phthalocyanines. Cells were incubated with TX-101A (●), TX 105A (▲), TX-106A (◆) or ZnTSPC (■) for 120 min. Bars represent SD of the mean of three experiments. (Reproduced with permission from *J. Porphyrins Phthalocyanines*, **2** (1998) 219, John Wiley & Sons Limited)

D. *Effect of Organic Compounds on the Aggregation of Zinc Phthalocyanines*

Zinc phthalocyanines easily aggregate in water and the aggregated phthalocyanines dissociate in the presence of detergents. Fig. 5.21 shows an example of the absorption spectrum of TX-101A. The absorption spectrum of TX-101A in water shows a strong absorption band at 653 nm corresponding to the aggregates. The absorption at 686 nm attributed to the monomeric form of TX-101A increases by the addition of the detergent, Tween 20. Table 5.2 shows the absorption maxima of some phthalocyanines in water and 1%Tween 20 aq. solution. All the phthalocyanines aggregate in water and dissociate in Tween 20 aq. solution.

The fluorescence spectrum of TX-101A is shown in Fig. 5.22. The fluorescence intensity increases with increase in Tween 20 concentration, showing that the fluorescence intensity of aggregated TX-101A is low. Table 5.3 shows the fluorescence intensities of some phthalocyanines

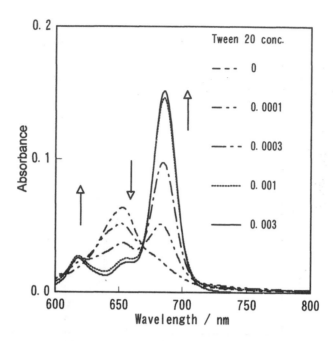

Fig. 5.21 Absorption spectra of TX-101A in surfactant solution. (Reproduced with permission from *J. Porphyrins Phthalocyanines*, **3** (1999) 1, John Wiley & Sons Limited)

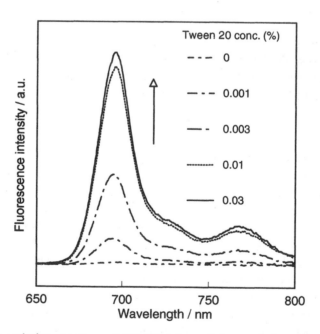

Fig. 5.22 Fluorescence emission spectrum of TX-101A in surfactant solution. (Reproduced with permission from *J. Porphyrins Phthalocyanines*, **3** (1999) 1, John Wiley & Sons Limited)

Table 5.3   Relative fluorescence intensity of phthalocyanine in various concentration of surfactant

	%Tween 20							
	0	0.001	0.003	0.01	0.03	0.1	0.3	1
TSPC	0.17*				0.60	0.82	1.00	1
MSPC	0.04				0.41	0.72	0.93	1
TX-101A	0.02	0.13	0.42	0.87	1			1
TX-105A	0.09	0.62	0.76	0.87	1			1
TX-106A	0.01	0.20	0.54	0.78	1			1

*: Fluorescence intensities were normalized to 1 obtained at 1 % Tween 20 solution. (Reproduced with permission from *J. Porphyrins Phthalocyanines*, **3** (1999) 1, John Wiley & Sons Limited)

in water and 1% Tween 20 aq. solution. These phthalocyanines in water show low fluorescence intensities, which increase in Tween 20 aq. solution. Thus, the fluorescence intensity of these phthalocyanines is attributed to their monomeric form.

In PDT studies, the cytotoxic species are formed by energy transfer from the photoexcited triplet state of the sensitizer to oxygen. Therefore, the triplet state of the sensitizer is significantly related to photodynamic efficiency. Fig. 5.23 shows the time profiles of the triplet state of TX-

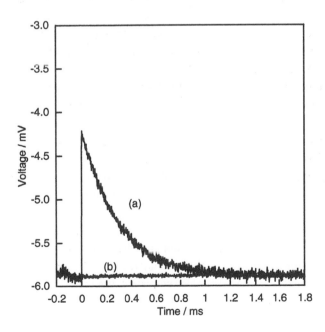

Fig. 5.23 Decay curve of excited triplet of TX-101A in 1% Tween 20 solution (a) and water (b). (Reproduced with permission from *J. Porphyrins Phthalocyanines*, **3** (1999) 1, John Wiley & Sons Limited)

101A (monitored at 470 nm) in water and 1% Tween 20 aq. solution. In Tween 20 aq. solution the photoexcited triplet state lifetime of TX-101A is estimated to be 320 ms. In water transient absorption of phthalocyanines is not detected because the intermolecular quenching of triplet state occurs in the aggregated phthalocyanine, indicating that the lifetime of the triplet state of aggregated phthalocyanine is very short. These aggregated phthalocyanines do not take part in the photodynamic action because of the short lifetime of the photoexcited triplet state.

The aggregation degree of phthalocyanine is affected by the surrounding conditions, especially organic solvents such as pyridine, DMF and ethanol. Fig. 5.24 shows the absorption spectra of zinc phthalocyanines in various organic solvents. The absorption spectrum of TX-101A in hydrophilic solvents such as water and methanol showed a strong absorption peak around 650 nm corresponding to the aggregates. In hydrophobic surroundings, the absorption of 670 nm of the monomeric form of TX-101A increases. In propanol, the absorption at 670 nm is maximum. In the case of TX-106A, the absorption of the monomeric form is maximum in octanol. In the case of TX-105A, the absorption of the monomeric form increases with hydrophobicity. TSPc shows the monomeric form in methanol and TSPc does not dissolve in more hydrophobic solvents. The absorption of the monomeric form of MSPc increases in ethanol and does not change in more hydrophobic solvents. Thus the dissociation conditions of these phthalocyanines are different from each other.

The dissociation of phthalocyanine in tissue is investigated by measuring fluorescence. Tumor and skin are incubated in MEM containing phthalocyanine. The fluorescence intensities per phthalocyanine concentration in the tissue are calculated. Table 5.3 shows the ratio of fluorescence intensities between tumor and skin. The fluorescence intensities of all phthalocyanines in tumor are stronger than that in skin. From these results, tumors may be more hydrophobic than skin. This difference may be attributed to dissociation properties.

The fluorescence intensities of TX-101A and TX-16A in skin are very weak, and TX-101A and TX-106A selectively dissociate in tumor. TX-106A indicates the highest ratio of fluorescence intensity and TX-101A indicates the second highest ratio of fluorescence intensity. These phthalocyanines dissociated in LDL solution but did not dissociate in BSA or ghost solution. Therefore, photosensitizers such as TX-101A and TX-106A having dissociation property may have selective photodynamic activity in tumor. These results indicate that use of the aggregation property of photosensitizers may be effective for the development of photosensitizers that activate selectively in tumor.

This dissociation property is not related to their accumulation mechanism. Indeed, TX-101A and TX-106A accumulate more in tumor than in skin. Thus, the photosensitizer that selectively activates in tumor can selectively accumulate in tumor. Therefore, the photosensitizer that selectively accumulates in tumor may also activate in tumor.

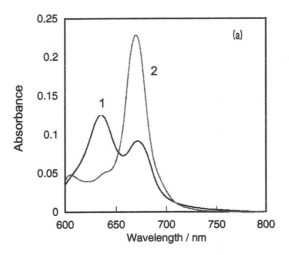

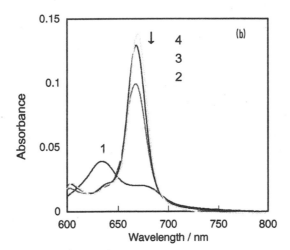

Fig. 5.24 Absorption spectra of phthalocyanine a: TSPC, b: MSPC, c: TX-101A, d: TX-105A, e: TX-101A. Absorption spectra of phthalocyanines were measured in water (1), methanol (2), ethanol (3), propanol (4), octanol (5), and decanol (6). (Reproduced with permission from *J. Porphyrins Phthalocyanines*, **3** (1999) 1, John Wiley & Sons Limited)

212

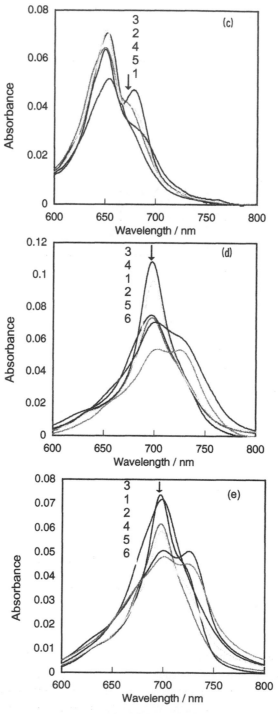

Fig. 5.24 continued.

E. Effects of Biological Materials on the Aggregation of Zinc Phthalocyanines

Table 5.4 shows the effect of biological material for aggregation of these phthalocyanines. The absorption spectra of these phthalocyanines are measured in the solution containing the biological material, e.g., BSA, LDL and ghosts. TSPc and MSPc dissociate upon addition of BSA. However, other phthalocyanines do not dissociate with the addition of BSA. TSPc and MSPc may bind to hydrophobic space in BSA. The absorbance of monomeric TX-101A and TX-106A increases on addition of LDL. LDL containing phospholipid and apoprotein has hydrophobic space in the molecules. TX-101A and TX-106A dissociate with addition of LDL solution because LDL is a more hydrophobic molecule than BSA. TSPc easily dissociates in hydrophobic space, but TSPc is the most hydrophilic of these phthalocyanines and does not dissolve in hydrophobic solution. Therefore, TSPc cannot penetrate in hydrophobic space of LDL. By the addition of ghosts, only MSPc dissociates, because the hydrophobicity of ghosts lies between that of BSA and LDL. Thus, these phthalocyanines differ from one another in dissociation properties.

Table 5.4 The efficiencies of biomaterial for formation of phthalocyanine monomer

	Biological material		
	BSA[a]	LDL[b]	RG[c]
TSPC	dissociation	N.D.[d]	N.D.
MSPC	dissociation	dissociation	dissociation
TX-101A	N.D.	dissociation	N.D.
TX-105A	N.D.	N.D.	N.D.
TX-106A	N.D.	dissociation	N.D.

a: Absorption spectra measured in 10 mg/ml BSA solution

b: Absorption spectra measured in 70 μg/ml LDL solution

c: Absorption spectra measured in 10 μg/ml RG solution

d: N.D. The red shift of the absorption band not dete

(Reproduced with permission from *J. Porphyrins Phthalocyanines*, **3** (1999) 1, John Wiley & Sons Limited)

5.4   Fluorescent Compounds in Tumor Cells

Fluorescent compounds are contained in some tumors. In this section the fluorescent compounds from human abdominal cancer, solid tumor of mice and Hela cells are described and the formation mechanism of these compounds is discussed.

5.4.1 Fluorescence from human abdominal cancer[25]

Figure 5.25a (see the color frontispiece) is a picture of human abdominal cancer (the arrows indicate the cancerous tissues), and Fig. 5.25b is the same picture with a cut filter of 600 nm.

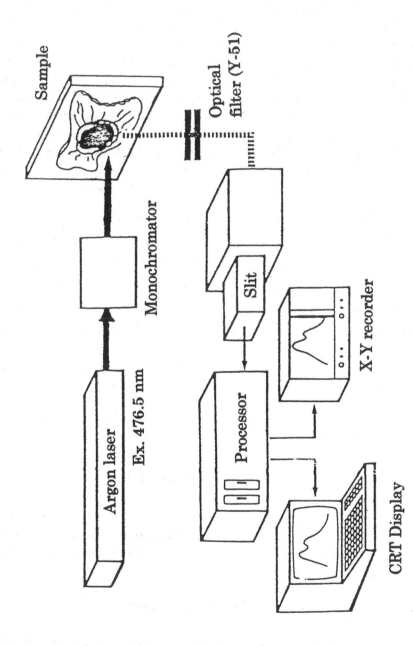

Fig. 5. 26 Detection system for laser induced fluorescence of gastro-intestinal cancer.

Orange-colored fluorescence is observed from the cancerous tissue. The apparatus to measure the fluorescence spectra of the cancer is shown in Fig. 5.26. An argon laser as the light source is used with an excitation wavelength of 476 nm. The fluorescence spectra are shown in Fig. 5.27. The number of each spectrum corresponds to the detected place, points, on the tumor surface, as shown in the figure. From the active tumor cells (No. 4) a fluorescence spectrum with a maximum wavelength of 620 nm is obtained; this is very similar to the fluorescence spectrum from the metal-free porphyrins. By irradiation with laser beam the intensity of the fluorescence decreases very rapidly. This kind of fluorescence is often observed in abdominal and esophagus cancer.

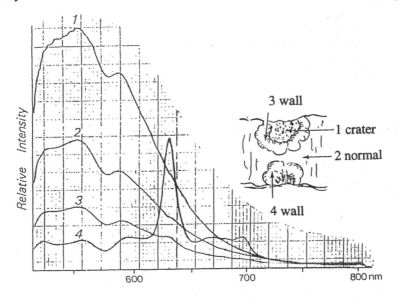

Fig. 5. 27 Fluorescence spectra from human abdoman cancer. (Reproduced with permission from M. Esaki et al., *Gastroenterol. Endoscopy*, **25** (1983) 56)

5.4.2 Fluorescence from solid tumor of mice[26]

Fluorescence is also detected from solid tumor of mice. The experimental procedure is shown in Fig. 5.28. C3H mice are used. MH-134 cancer cells are implanted and after 10 days the tumors are removed. After homogenization and centrifugation of the tumor in a mixed solvent the liquid phase is analyzed.

Comparing the fluorescence spectrum of the tumor extract and those of the reference solutions, it was found that the tumor extract contained zinc and metal-free protoporphyrins.

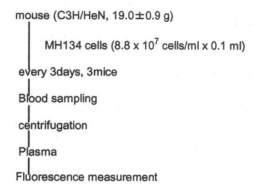

Fig. 5.28 Pretreatment of blood of mice for fluorescence measurement.

5.4.3 Fluorescence from blood of tumor-implanted mouse[26,27]

The tumor cells are injected under the skin of C3H mice. Every three days blood is collected from three mice. The blood is centrifuged and the plasma analyzed by spectrometer.

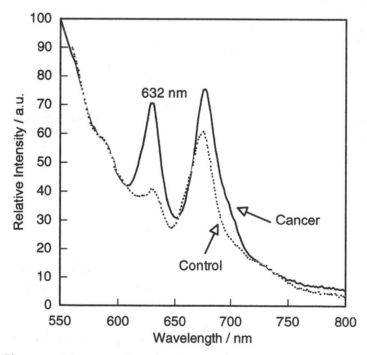

Fig. 5.29 Fluorescence spectrum from the plasma of the tumor implanted mouse. λ_{EX} = 390 nm
(Reproduced with permission from *Chem. Lett.*, (**1995**) 325)

A typical fluorescence spectrum from the plasma of the tumor-implanted mouse is shown by the solid line of Fig. 5.29. The dotted line is the reference (blood from a normal mouse). To compare these spectra the fluorescence peak at 625 nm is characteristic of tumor mouse. The fluorescence intensity depends on the time after tumor implantation, as shown in Fig. 5.30.

The fluorescence intensity increases with the number of days after tumor transplantation, and then decreases. At around 20 days the tumors become big enough and the mice are near death. At this time the tumor is dying and the fluorescence intensity decreases. However, when the tumor is active, the fluorescence intensity increases. To measure the fluorescence intensity, the mice with or without tumors may be identified.

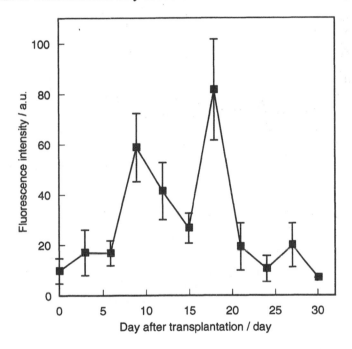

Fig. 5.30 Time dependence of fluorescence intensity of the blood of cancer implanted mouse. λ_{EX} = 390 nm, λ_{EM} = 632 nm. (Reproduced with permission from *Chem. Lett.*, (**1995**) 325)

5.4.4 Fluorescence from cultivated HeLa cells

In the cultivation medium of HeLa cells zinc protoporphyrin, metal-free protoporphyrin and coproporphyrin are observed. During the cultivation of HeLa cells these porphyrins increase as shown in Fig. 5.31. At this time the cell density increases from 10^5 to 10^6.

Besides HeLa cells, such porphyrin formation is also observed in Hepa II cells (a kind of

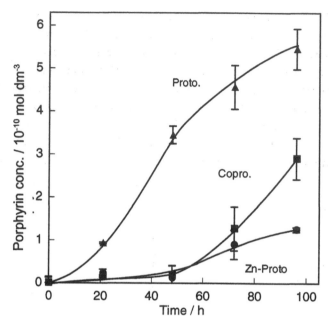

Fig 5.31   Time dependence of porphyrin formation during HeLa cell cultivation.

Table 5.5   Porphyrin production by various cell strains*

Strain	Porphyrin conc. / 10^{-10} mol dm^{-3}		
	Coproporphyrin I	Zinc Protoporphyrin IX	Protoporphyrin IX
HeLa	1.3±0.5	0.9±0.3	4.6±0.6
Hepatoma	0.4±0.2	0.5±0.4	1.5±0.1
Fibroblast (bone marrow)	0.6±0.0	0.0±0.1	0.1±0.1
Fibroblast	0.2±0.3	0.1±0.1	0.0±0.0

* The cells were cultured in an area of 21 cm²/dish with 4.0 ml supplemented MEM. 5.0 x 10⁵ cells were placed into the dish.   After 3 days cultivation, the porphyrin concentration was measured by HPLC.   Results represent the mean ± SD.   (Reproduced with permission from *Porphyrins*, **6** (1997) 15)

cancer cell), as shown in Table 5.5.   Such porphyrin formation is not observed in fibroblasts (normal cells).   In cancer cells, the concentration of zinc and metal-free protoporphyrins increases, but in normal cells zinc and metal-free protoporphyrins are not formed, indicating that the formation of these porphyrins may be specific to cancer cells.

5.4.5 Reaction mechanism of the formation of protoporphyrin and copro-porphyrin

Figure 5.32 shows the biological pathway of heme protein formation. In normal cells protoporphyrin is an intermediate compound, for the rate of heme formation from protoporphyrin is very high. In normal cells no accumulation of protoporphyrin or coproporphyrin is observed. Once the enzyme coproporphyrinogen synthase is partially inhibited, another reaction proceeds and coproporphyrin is formed. Since coproporphyrin is detected in cancer cells, coproporphyrin synthase is partially inhibited. Zinc and metal-free protoporphyins accumulate in cancer cells. One possible reason for this is the inactivation of the enzyme of the reaction from protoporphyrin

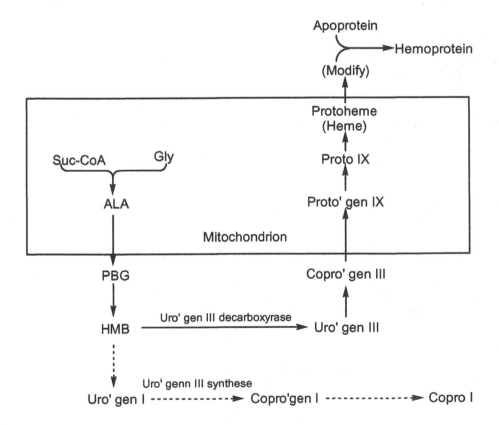

Fig. 5.32 Biological pathway of heme protein formation.

to heme, resulting in the accumulation of protoporphyrin. Another possibility is the existence of the reverse reaction, demetallation; iron is removed from the heme and protoporphyin is formed as shown in Fig. 5.33. When protoporphyrin accumulates zinc ions are easily inserted into the porphyrin ring nonenzymatically.

When hemoglobin or hemin is introduced in HeLa cell cultivation and the protoporphyin and zinc protoporphyin in the medium solution analyzed, zinc and metal-free protoporphyrins are formed as shown in Table 5.6. This indicates the existence of the reaction of demetallation.

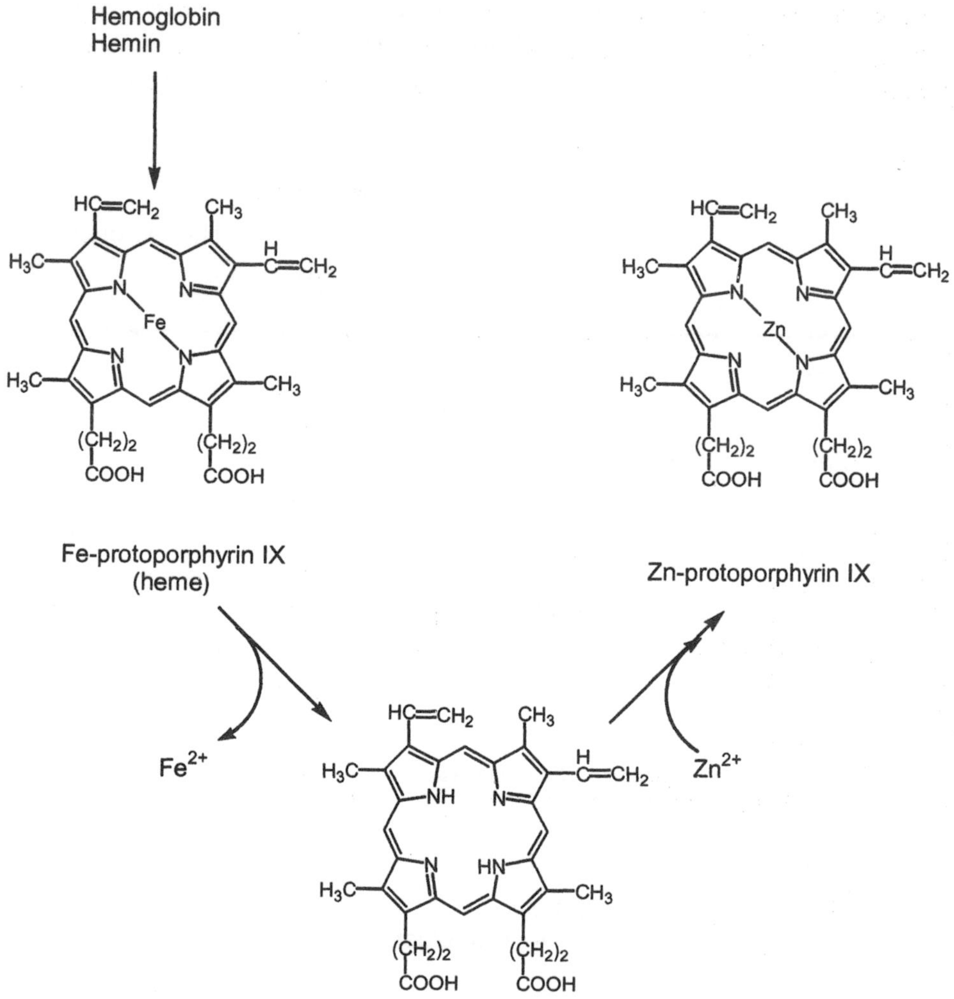

Fig. 5.33 Proposed reaction pathway of demetallation of iron ion and insertion of zinc ion.

Table 5.6   Effect of hemoglobin and hemin for porphyrin formation

| | Porphyrin conc. /10^{-10} mol dm^{-3} | | |
	Copro	Zn-Proto	Proto
None	4.4	0	0
Hemoglobin	-	0.3	66.6
Hemin	0.6	0.9	84.7

(Reproduced with permission from *Chem. Lett.*, (**1995**) 325)

5.5   5-Aminolevulinic Acid (ALA)

Recently, there has been considerable interest in a new approach to photodynamic therapy in which a nonphototoxic prodrug, 5-Aminolevulinic acid (ALA) is administered, resulting in the endogeneous synthesis of the photosensitizer, protoporphyrin (PP).[28-32] The addition of ALA is a rate-limiting step in the heme biosynthetic pathway; increase in endogenous porphyrin can be artificially induced by an exogenous supply of excess ALA. There are several possible mechanisms that can enhance the concentration of PP in tumor compared with that in normal tissue, e.g., through having a different heme synthesis profile or a relative deficiency in the ferrochelatase, converting PP to heme. It is known that the last steps in the heme synthetic pathway, including PP synthesis, occur in mitochondria, and that PP is induced by ALA accumulated in mitochondria, as shown in Fig. 5.34.

Thus, the accumulation mechanism of ALA-induced PP is apparently different from that of other exogenous sensitizers. The photodynamic efficiencies of exogenous sensitizers strongly depend on the type of sensitizer. Therefore, the photodynamic efficiencies of ALA-induced PP may be different from that of exogenous PP. In this section a comparison of the photodynamic efficiency of exogenous PP and ALA-induced PP is described.

PP is accumulated by two protocols. One protocol involves accumulation after the incubation of HeLa cells with 1×10^{-6} mol dm^{-3} PP, and the other involves formation of PP after incubation of HeLa cells with 7.0×10^{-4} mol dm^{-3} ALA.

Figure 5.35 shows the time dependence of PP accumulation in HeLa cells. The HeLa cells are incubated with 7.0×10^{-4} mol dm^{-3} ALA at 37 °C (Fig. 5.35a). Intracellular PP concentration increases after the induction period. When HeLa cells are incubated with 1×10^{-6} mol dm^{-3} PP at 37°C, PP accumulates rapidly, as shown in Fig. 5.35b. Fig. 5.36 shows percent survival after irradiation as a function of incubation time. When incubated with ALA, percent survival decreases rapidly after a 0.5-h incubation period, as shown in Fig. 5.36a. When HeLa cells are incubated with PP percent survival decreases gradually with incubation time. Fig.5.37

222

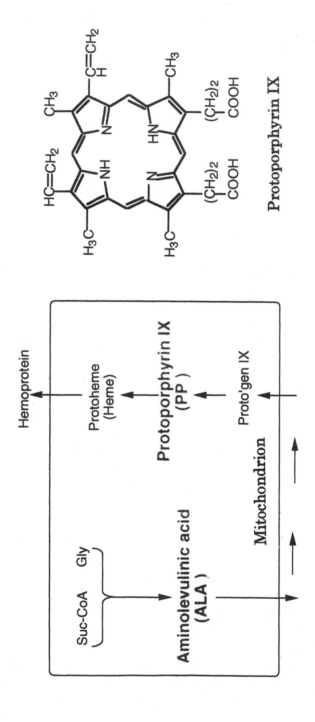

Fig 5.34 Reaction pathway of protoporphyrin IX formation from precursor, aminolevulinic acid (ALA).

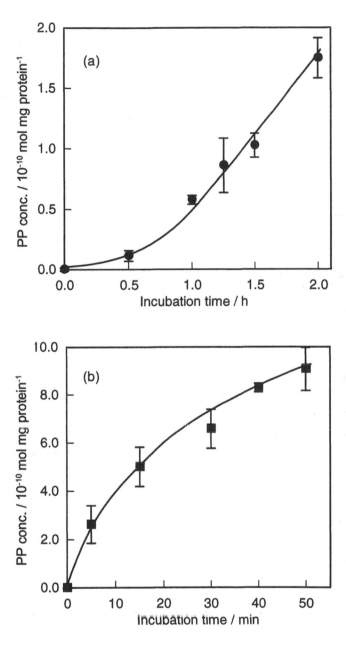

Fig. 5.35 Time dependence of PP accumulation in HeLa cells. HeLa cells were incubated with 2 ml of 7.0×10^{-4} mol dm^{-3} ALA (a) or 1.0×10^{-6} mol dm^{-3} PP (b). At each point, intracellular PP was measured. Each point represents the mean ± SD. (Reproduced with permission from *Photochem. Photobiol.*, **66** (1997) 842, American Society for Photobiology)

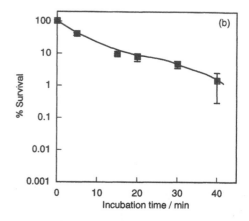

Fig. 5.36 Survival of HeLa cells as a function of incubation time after exposure for 8 min to red light (λ > 500 nm; 65 mW / cm^2). HeLa cells were incubated with 2 ml of 7.0×10^{-4} mol dm^{-3} ALA (a) or 1.0×10^{-6} mol dm^{-3} PP (b). At each point, HeLa cells were irradiated. Each point represents the mean ± SD. (Reproduced with permission from *Photochem. Photobiol.*, **66** (1997) 842, American Society for Photobiology)

shows the survival of HeLa cells after irradiation as a function of intracellular PP concentration. With either protocol, photodynamic efficiency increases with intracellular PP concentration. However, endogenous PP (LD90 = 1.2×10^{-10} mol/mg protein) is more efficient than exogenous PP (LD90 = 4.5×10^{-10} mol/mg protein).

The results may be explained in two ways. One reason is that the intracellular PP monomer concentration is different for each protocol. Porphyrin monomer concentration is an important factor in photodynamic efficiency, because aggregated porphyrin causes low photodynamic action. The other reason is the difference in the site of cell damage.

As aggregated porphyrin shows very low fluorescence quantum yield, the fluorescence intensity of the cell may be related to the PP monomer concentration. Fig. 5.38 shows the fluorescence intensity of HeLa cells as a function of intercellular PP concentration. Since the fluorescence intensity of endogenous PP is higher than that of exogenous PP, exogenous PP may aggregate in the cell. Fig. 5.39 shows the survival of HeLa cells as a function of fluorescence intensity. The difference in photodynamic efficiency cannot be explained only by intracellular PP monomer concentration. Therefore, the site of cell damage may differ between endogenous PP and exogenous PP.

The damage to mitochondria and plasma membrane has been measured and is considered to be related to cell death. The site of cell damage is compared by measuring LDH leakage and mitochondrial enzyme activity. To investigate plasma membrane damage, LDH leakage is measured (Fig. 5.40). Fig. 5.40a shows the time dependence of LDH leakage. After irradiation, LDH loss is observed in both protocols. Leakage of LDH induced by exogenous PP occurs earlier when

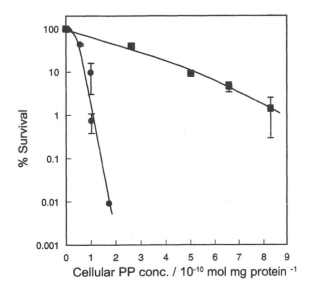

Fig. 5.37 Survival of HeLa cells as a function of intracellular PP concentration after exposure for 8 min to red light ($\lambda > 500$ nm; 65 mW / cm²). HeLa cells were incubated with 7.0×10^{-4} mol dm⁻³ ALA (●) or 1.0×10^{-6} mol dm⁻³ PP(■). Each point represents the mean ± SD. (Reproduced with permission from *Photochem. Photobiol.*, **66** (1997) 842, American Society for Photobiology)

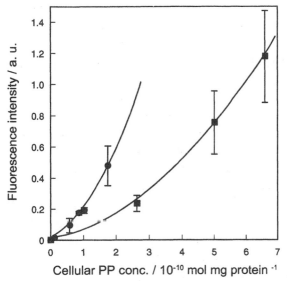

Fig. 5.38 Fluorescence intensity as a function of intracellular PP concentration. HeLa cells were incubated with 7.0×10^{-4} mol dm⁻³ ALA (●) or 1.0×10^{-6} mol dm⁻³ PP(■). Each point represents the mean ± SD.

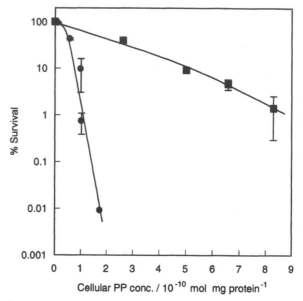

Fig. 5.39 Survival of HeLa cells as a function of fluorescence intensity of intracellular PP after exposure
for 8 min to red light (λ > 500 nm; 65 mW / cm^2). HeLa cells were incubated with 2 ml of
7.0 x 10^{-4} mol dm^{-3} ALA (●) or 1.0 x 10^{-6} mol dm^{-3} PP (■). Each point represents the
mean ± SD.

induced by endogenous PP. Fig. 5.40b shows LDH leakage 3 h after irradiation. At the same
survival rate, exogenous PP causes more LDH leakage than endogenous PP. Thus, exogenous
PP may preferentially cause photodamage to the plasma membrane.

Figure 5.41 shows 4.5-dimethyl thiazol-2-yl 2, 5-diphenyl tetrazolium bromide(MTT)
reduction activity as a function of intracellular PP. With either protocol, MTT reduction activity
decreases with intracellular PP concentration. Endogenous PP inhibits MTT reduction activity
more strongly than exogenous PP. Reduction of triphenyl-tetrazolium chloride(TTC) activity
also decreases the same as MTT. Reduction rates of MTT and TTC are related to mitochondrial
enzyme activity. These results indicate that endogenous PP selectively sensitizes mitochondrial
functions. Photodynamic damage occurs around the site of the photosensitizer adsorbed in the
cell, because $^1O_2$ can diffuse significantly in cells during its lifetime. Therefore, endogenous PP
may selectively localize in mitochondria while the data indicate that exogenous PP localize mainly
in the plasma membrane. The ALA-induced PP accumulation in mitochondria has been reported
by investigation using confocal microscopy, and intracellular localization of the porphyrin
supplemented in the medium has been reported to occur first within the plasma membrane.

Thus the difference in photodynamic efficiency of intracellular monomeric PP is caused
by a difference in the PP localization site and in the resulting cell damage. Molecules of exogenously
supplied PP are less phototoxic than PP generated endogenously from ALA.

The explanation appears to be related to the sites of photodamage, with exogenous PP

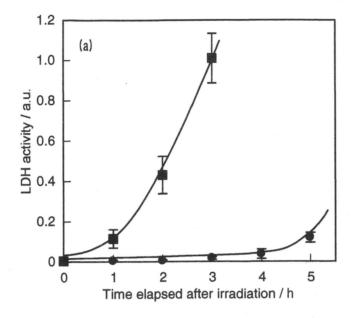

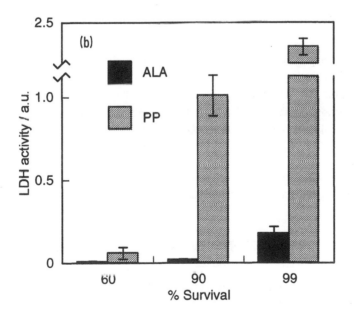

Fig. 5.40 LDH leakage as a function of time elapsed after exposure for 8 min to red light ($\lambda > 500$ nm; 65 mW / cm^2) (a). HeLa cells were incubated with 2 ml of 7×10^{-4} mol dm^{-3} ALA(■) or 1×10^{-6} mol dm^{-3} PP (●). LDH leakage after 3 h irradiation (b). Each point represents the mean ± SD.

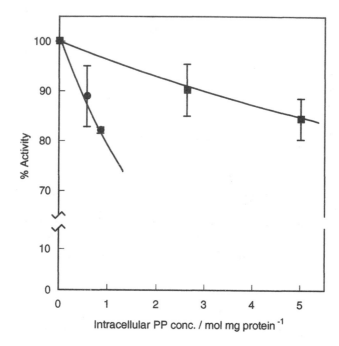

Fig. 5.41 MTT reduction activity as a function of intracellular PP concentration after exposure for
8 min to red light (λ> 500 nm; 65 mW / cm²). HeLa cells were incubated with 2 ml of 7.0
x 10⁻⁴ mol dm⁻³ ALA (●) or 1.0 x 10⁻⁶ mol dm⁻³ PP (■). After irradiation, HeLa cells
were incubated with 10 mg / ml MTT. Each point represents the mean ± SD. (Reproduced
with permission from *Photochem. Photobiol.*, **66** (1997) 842)

affecting membrane permeability functions while endogenous PP causes photodamage to
mitochondrial enzymes, suggesting mitochondrial photodamage.

5.6   Antibody-photosensitizer Conjugate

For selective accumulation of a photodrug in tumor tissues, conjugation of the photodrug to
biomolecules such as monoclonal antibodies has been studied.[33-45] Malignant tumor cells have
cell surface antigens which are different from those of normal cells. Monoclonal antibodies
directed toward tumor cell antigens are able to couple with photosensitizers. The antibody-
photosensitizer conjugate will bind specifically to a tumor tissue and induce photokilling without
damaging normal tissues. This difference in tumor and normal uptake is favorable for

immunotherapy of monoclonal antibody-photosensitizer conjugates. As an example, antibody-photosensitizer conjugate for mouse MM 46 tumor cells is described.

The direct binding of a photosensitizer to the constant part of monoclonal antibodies requires the presence of a suitable chemical group in the photosensitizer.Tetra-carboxylated zinc-phthalocyanine(ZnTCPc) is coupled to IgG-antibodies by a method using N-hydroxysuccinimide as the linking agent shown in Scheme 5-2. About 20 molecules of phthalocyanine bind covalently to one antibody molecule.

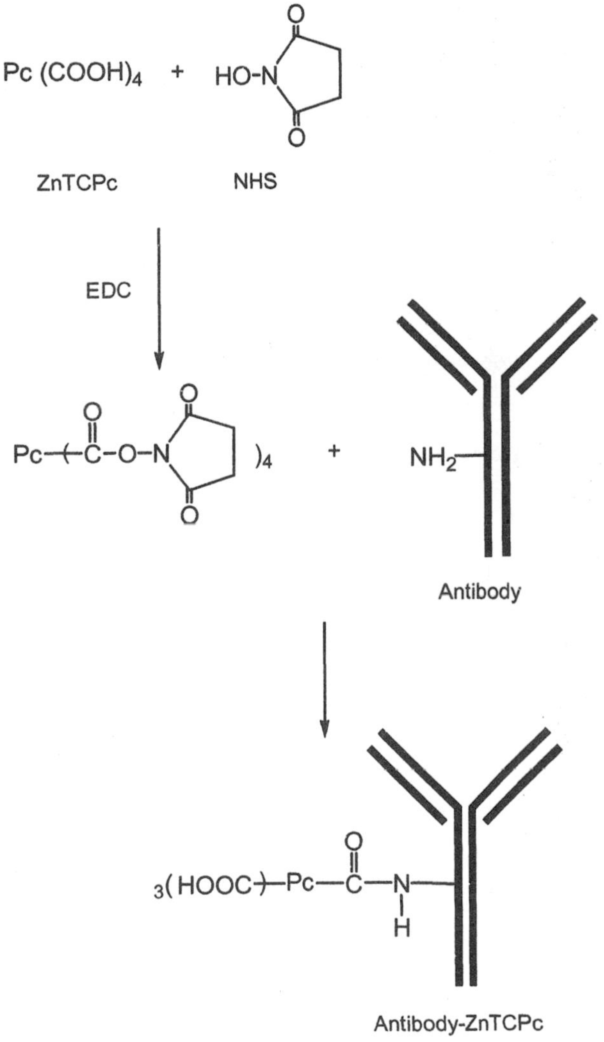

Scheme 5-2 Synthesis of phthalocyanine-immunoconjugate.

The cellular uptake of the antibody-photosensitizer conjugate is shown in Fig. 5.42. The conjugate is selectively incorporated into the tumor tissue. The conjugate-incorporated tumor tissues are strongly damaged by irradiation with laser beam, as shown in Fig. 5.43. This technique enables the selective delivery of a photodrug to tumor tissue.

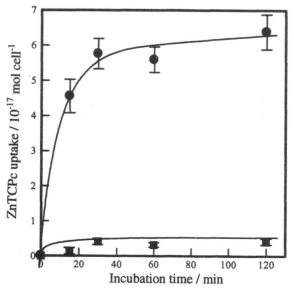

Fig. 5.42 Cellular uptake of ZnTCPc (●), IgG-ZnTCPc(■).

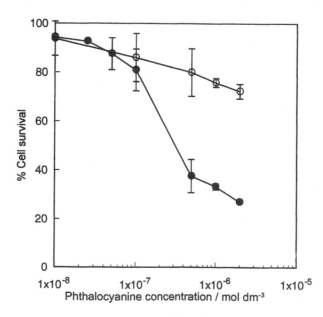

Fig. 5.43 Photocytotoxity of ZnTCPc (O), IgG-ZnTCPc(●). MM46 cells were incubated for 2h.

References

1. I. Diamond, S. Granelli, A. F. McDonaph, S. Nielsen, C. B. Wilson and R. Jaenkcke, *Lancet*, **2**, 1175-77 (1972)
2. T. J. Dougherty, G. Lawrence, J. E. Kaufman, D. Boyle, K. R. Weishaupt and A. Goldfarb, *J. Natl. Cancer Inst.*, **62**, 231-7 (1979)
3. T. J. Dougherty, G. Lawrence, J. E. Kaufman, A. Goldfarb, K. R. Weishaupt and Mittelman, *Cancer Res.*, **38**, 2628-35 (1978)
4. G. Jori, *Photochem. Photobiol.*, **52**, 439-43 (1990)
5. R. Bonnett, *Chem. Soc. Rev.,* **24**, 19-23 (1995)
6. R. E. Lipson, E. Baldes and A. M. Olsen, *J. Natl. Cancer Inst.*, **26**, 1-11 (1961)
7. R. Bonnett, *Chem. Soc. Rev.*, 19-33 (**1995**)
8. S. Wan, J. A. Parrish, R. R. Anderson and M. Madden, *Photochem. Photobiol.*, **34**, 679-81 (1981)
9. D. J. McGarvey, T. G. Truscott in *Photodynamic Therapy of Neoplastic Disease* (ed. D. Kessel), vol. 2, pp. 179-89, CTC Press, Florida (1990)
10. T. Nishisaka, N. Sugiyama, I. Okura, *Laser in the Life Science*, **4**, 87-92 (1991)
11. M. Toriya, S. Yaginuma, S. Murofushi, K. Ogawa, N. Muto, M. Hayashi, and K. Matsumoto, *J. Antibiotics*, **46**, 196-200 (1993)
12. L. I. Gossweiner, in *Porphyric Pesticides* (ed. by S. O. Duke and C. A. Rebeiz) pp. 255-265. ACS Symposium Series 559, American Chemical Society, Washington, D.C. (1994)
13. B. Paquette, H. Ali, R. Langlis. and J. E. van Lier, *Photochem. Photobiol.* **47**, 215-220(1998)
14. T. Takeno, J. Hirota, K. Kamachi, I. Okura, Y. Sakuma and T. Nishisaka, *Lasers in the Life Sciences,* **6**, 173-179 (1994)
15. D. Wohrle, N. Iskander, G. Graschew, H. Sinn, E. A. Friedrich. W. Maier-Borst. J. Stern and P. Schlag, *Photochem. Photobiol.,* **51**, 351-356 (1990)
16. U. Michelsen, H. Kliesch, G. Schnurpfeil, A. K. Sobbi, D. Wohrle, *Photochem. Photobiol.*, **64**, 694-701 (1996)
17. K. Fukushima, K. Tabata and I. Okura, *J. Porphyrins Phthalocyanines*, **2**, 219-222 (1998)
18. C. C. Leznoff, S Vigh P. I. Svirskaya, S Greenberg D. M. Drew, E. Ben-Hur and I. Rosenthal, *Photochem. Photobiol.*, **49**, 279-284 (1989)
19. W. S. Chan, J. F. Marshall, R. Svensen, J, Bedwell and I. R. Hart, *Cancer Res.*, **50**, 4533-4538 (1990)
20. N. Ishikawa, O. Ohno, Y. Kaizu and H. Kobayashi, *J. Phys. Chem.*, **96**, 8832-8839 (1992)
21. D. Wohrle, A. Wendt, A. Weitemeyer, J. Stark, W. Spiller, S. Muller, U. Michelsen, H. Kliesch, A. Heuermann and A. Ardeschirpur, *Russian Chemical Bulletin,* **43**, 1953-1964 (1994)
22. J. R. Phillippot, A. G. CooPer and D. F. H. Wallach, *Proc. Natl. Acad. Sci. USA*, **74**, 956-

960 (1977)

23. I. Freitas, *J. Photochem. Photobiol.*, **7**, 359-369 (1990)

24. C. J. Welsh, M. Robinson, T. R. Warne, J. H. Rierce, G. C. Yeh and J. Phang, *Arch. Biochem. Biophys.*, **315**, 41-47 (1994)

25. M. Esaki, T. Nishisaka and M. Yonekawa *Gastroenterol. Endoscopy.*, **25**, 56-63 (1983)

26. T. Nishisaka, T. Fukami. K. Tabata, D. Wohrle and I. Okura, *Chem. Lett.*, **4**, 325-326 (1995)

27. T. Sugimura, M, Ueda and T. Ono, *Gann*, *47*, 87-90 (1956)

28. J. C. Kennedy, R. H. Pottier, J. Photochem. Photobiol., B-14, 245-292 (1992)

29. N. Rebiez, C. C. Rebiez, S. Arkins, K.W. Kelly, C. A. Rebiez, *Photochem. Photobiol.*, **55**, 431-435 (1992)

30. M. R. Moore, K. E. L. McColl, C. Rimington and A. Goldberg: in *Disorders of Porphyrin Metabolism* (Eds. by M. R. Moore, K. E. L, McColl, C. Rimington and A. Goldberg), pp. 21-72. Plenum Press, NY. (1987)

31. B. C. Wilson and G. Sngh, *Photochem. Photobiol.*, 65: 166-176 (1997)

32. K. Tabata, S. Ogura and I. Okura, submitted

33. A. Bamias. P. Keane, T. Krausz. G. Williams and A. A. Epenetos, *Cancer Research*, **51**, 724-728 (1991) .

34. T. A Dahl, *The Spectrum*, 11-16 (**1992**)

35. A. R. Oseroff. D. Ohuoha, T. Hasan, J. C. Bommer, M. L. Yarmush, *Proc. Natl. Acad. Sci. (USA).*, **83**, 8744-8748 (1986)

36. J. Morgan, H. Lotman, C. C. Abbou and D. K. Chopin, *Photochem. Photobiol.*, **60**, 486-496 (1994)

37. D. Mew, C. K. Wat, G. H. N. Towers and J. G. Levey, *J. Immunol.*, **130**, 1473-1477 (1983)

38. G. A Goff. M. Bamberg and T. Hasan, *Cancer Res.*, **51**, 4762-4767 (1991)

39. J. W. Dawson, J. Morgan, H. Barr, B. R. Davidson, A J. MacRobert, F. Savage, C. Dean and P. B. Boulos, *Photochem. Photobiol.*, **53**, 25S (1991)

40. F. N. Jiang, *J. Contr. Release*, **19**, 41 (1992)

41. A. M. Richter, B. Kelly, J. Chow, D. J. Lui, G. H. N. Towers, D. Dolphin and J. G. Levy, *J. Natl. Cancer Inst.*, **79**, 1327 (1987)

42. G. Jori, *J. Photochem. Photobiol. A: Chem*, **62**, 371-378 (1992)

43. S. Devanathan, T. A Dahl, W. R. Midden, D. C. Neckers, *Proc. Natl. Acad. Sci (USA)*, **87**, 2980 (1990)

44. B. Bartik, A. Weitemeyer, S. Kaul, D. Wohrle unpublished results

45. B. C. Schultes, S. Spaniol, S. Popat. W. Ertmer and S. Schmidt, Elsevier Science Publishers b. V. *Photodynamic Therapy and Biomedical Lasers* (P. Spinelli, M. D. Fante and R. Marchesini, eds.), 996, 333-337 (1992)

Appendix

As described in Chapter 3, effective photoinduced charge separation is one of the most promising approaches to the use of solar energy, which enables the supply of reducing power. And the utilization of solar energy is remarkably expanding.

As an application of reducing power caused by sunlight, it may help to solve serious environmental problems. The international agenda requires not only the creation of new renewable energy resources, but also solutions for serious global environmental problems.

The process of photolysis of water would greatly impact the refinement of industrial technology and start a new trend in the production of alternate energy source, hydrogen, the combustion of which produces water. Recycling of water would provide almost unlimited energy without causing environmental pollution. Toxic gas emissions resulting from burning of fossil fuels and from fermentation of organic compounds which increase the levels of carbon dioxide and methane causing global warming and climatic changes are effectively controlled by fixation by the methanogenesis process. This process would also yield a useful product of economic importance, methanol.

The process converting carbon dioxide and methane into methanol is shown in Scheme I, which is a combination of three systems. This process has the multiple advantages of (1) producing an alternate source of fuel, hydrogen gas, (2) converting the excess atmospheric pollutant, carbon dioxide, into methane, and (3) into methanol, an economically useful industrial solvent by the partial oxidation of methane.

1. Photolysis of water with solar energy

As shown in Scheme 3-4 (the brief reaction mechanism of the green plant photosystem), green plants split water under irradiation by sunlight. Green plants extract electrons by water oxidation with the aid of light energy and the electrons are transported into the Calvin cycle. By the addition of hydrogenase to extract electrons from the electron flow, hydrogen evolves through

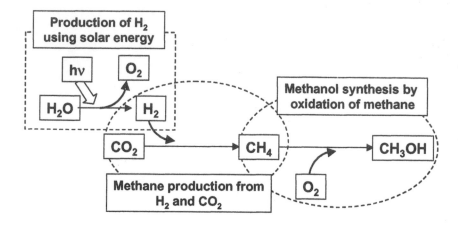

Scheme I Schematic diagram of methanol synthesis by using carbon dioxide and methane.

the reduction of protons.

Another process is the combination of oxygen evolution and hydrogen evolution systems shown in Scheme II. The system contains grana, electron carrier such as NAD and NADP, and hydrogenase.

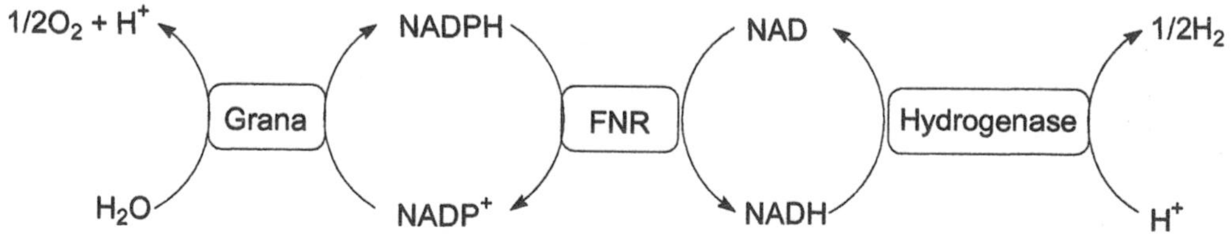

$1/2O_2 + H^+$ ... NADPH ... NAD ... $1/2H_2$

Grana   FNR   Hydrogenase

H_2O ... $NADP^+$ ... NADH ... H^+

Scheme II   Combination of oxygen evolution and hydrogen evolution system. FNR: Ferredoxin-NADP$^+$ oxidoreductase.

2. Conversion of carbon dioxide into methane with *Methanobacterium*

Methanobacterium can produce methane by the reduction of carbon dioxide with hydrogen.

$$CO_2 + 4H_2O \rightarrow CH_4 + 2H_2O$$

Methanobacterium thermoautotrophicum is one of the most effective methane forming bacterium.

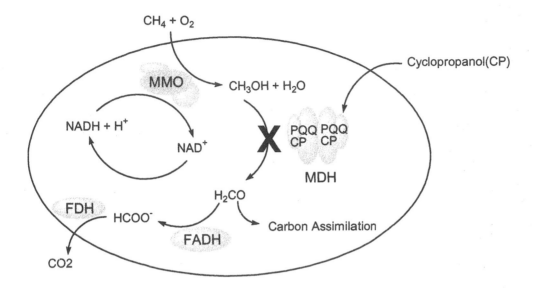

Scheme III   Pathway of methanol sysnthesis using *Methylosinus trichosporium* OB3b. MMO: Methane
monooxygenase, MDH: Methanol dehydrogenase, FADH: Formaldehyde dehydrogenase,
FDH: Formate dehydrogenase, PQQ: Pyrroquinoline quinone.

3. Conversion of methane into methanol with methanotrophs

Methane-oxidizing bacteria (Methanotrophs) are unique in their ability to utilize methane as the
sole carbon source. As shown in Scheme III, the pathway has been advanced for methane
metabolism by methanotrophs. Methanotrophs are aerobic bacteria that convert methane to
methanol in the first step of the metabolic pathway and utilize molecular oxygen for the step. A
unique monooxygenase, methane monooxygenase, catalyzes the hydroxylation of methane to
methanol according to the following equation.

$$CH_4 + O_2 + 2e + 2H^+ \rightarrow CH_3OH + H_2O$$

However, the methanol production is subsequently oxidized by methanol dehydrogenase in the
same bacterium and completely oxidized to carbon dioxide in the end. When the methanol
dehydrogenase is selectively inhibited, methanol is accumulated. Since methanol dehydrogenase
contains pyrroloquinoline quinone (PQQ) as a cofactor, cyclopropanol can react PQQ selectively
by the ring opening of cyclopropanol. When the methanotrophs are treated with cyclopropanol,
they serve as a methanol production catalyst.

In this appendix the above process is described as an example having the multiple advantages of producing an economically useful industrial material, methanol. Many other processes are available as the result of photosensitization of porphyrins and phthalocyanines.

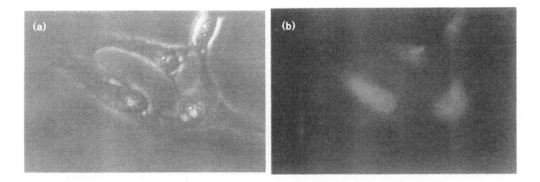

Fig. 4.24 Images of PdTCPP-stained HcLa cells taken using (a) phase contrast and (b) fluorescence modes. (see the text, p.176)

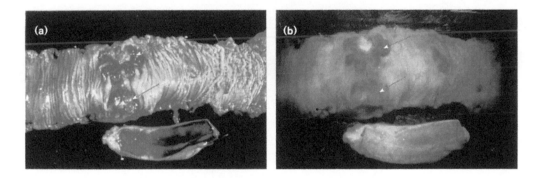

Fig. 5.25 (a) Picture of a human abdominal cancer. (b) the same of (a) with 600 nm cut filter. (see the text, p.213)

Abbreviations

ALA:	5-AminolevulinicAcid
AlPc(OH):	aluminum 2,9,16,23-tetraphenoxy-29H, 31H-phthalocyanine hydroxide
AlPcCl:	aluminium phthalocyanine chloride
BePc:	beryllium phthalocyanine
Bpy:	2,2'-bipyridine
BrCnBr:	a,w-dibromoalkane
C60:	fullerene
CaPc:	calcium phthalocyanine
CCD:	charge-coupled device
Cd-TPPS$_3^{3-}$:	cadmium tetraphenylporphyrin tetrasulfonate
Chl:	chlorophyll
CMC:	critical micellar concentration
Compound (A):	1,1'-dimethyl-2,2'-bipyridinium dichloride
Compound (B):	6,7-dihydrodipyrido[1,2-a:2,1-c] pyrazinium dichloride
Compound (C):	6,7-dihydro-6-H-dipyrido[1,2-a:2,1-c] pyrazinium dichloride
Compound (D):	6,7,8,9-tetrahydrodipyrido[1,2-a:2,1-c] -[1,4]-diazocinium dic
CoTPP:	cobalt tetraphenylporphyrin
Co-TPPS$_3^{3-}$:	cobalt tetraphenylporphyrin tetrasulfonate
C-P-Q:	quinone-linked porphyrins with carotenoid
C-P-QA-QB:	bis-quinone-linked porphyrins with carotenoid
C-PZn-P-QA-QB:	bis-quinone-linked bis-porphyrins with carotenoid
CTAB:	cethyltrimethylammonium bromide
Cu Pc(CH$_2$OC$_{12}$H$_{25}$)$_8$:	copper phthalocyanines substituted by long paraffinic ch
CuPc:	copper phthalocyanine
CuTPP:	copper tetraphenylporphyrin
Cu TPPS$_3^{3-}$:	copper tetraphenylporphyrin tetrasulfonate
DBU:	1,8-diazabicyclo[5,4,0]undensene
DDQ:	2,3-dichloro-5,6-dicyano-1,4-benzoquinone
DMF:	dimethylformamide
DMSO:	dimethylsufoxide
DO.D.:	optical density
E(P / P*):	energy of the photoexcited triplet state
E(P* / P):	one electron reduction potential
E(P$^+$ / P*):	one electron oxidation potential
ε:	molar absorption coefficient

EDTA:	ethylenediamine tetraacetic acid
EEP:	5-ethyl-etioporphyrin
ET:	energy of the excited triplet state
EtioP:	etioporphyrin
Φ:	quantum yield
FeOEP:	iron(III) octaethylporphyrin
FePc:	iron phthalocyanine
ferron:	8-hydroxy-7-iodo-5-quinolinesulfuric acid
Φ_F:	fluorescence quantum yield
FNR:	ferredoxin-NADP reductase
Φ_P:	phosphorescence quantum yield
Φ_T:	triplet yield
GP-197:	polydimethylsiloxane derivative
H_2ase:	hydrogenase
H_2TPP:	tetraphenylporphyrin
H_2-TPPS$_3{}^{3-}$:	tetraphenylporphyrin tetrasulfonate
HmP:	hematoporphyrin
HOMO:	highest occupied molecular orbital
HpD:	Hematoporphyrin derivative
Ip_0:	phosphorescence intensities in the presence of reaction substrate
Ip:	phosphorescence intensities in the absence of reaction substrate
I:	Total reaction rate
KSV:	Stern-Volmer constant
Ksv:	Stern-Volmer constant
λmax:	peak of the absorption band with maximum wavelength
LEDs:	light emitting diodes
Li_2Pc:	dilitium phthalocyanine
LUMO:	lowest unoccupied molecular orbital
m-PCnV:	meta-viologen-linked porphyrin
M:	reaction substrate (quencher)
MgPc(C_6H_5)$_4$:	magnesium tetraphenylphthalocyanine
MgPc(NO_2)$_4$:	magnesium tetranitorophthalocyanine
MgPc(OCH_3)$_4$:	magnesium tetramethoxyphthalocyanine
MgPc:	magnesium phthalocyanine
MgTPP:	magnesium tetraphenylporphyrin
MgTPPS	magnesium tetraphenylporphyrin tetrasulfonate
Mn-TPPS$_3{}^{3-}$:	manganese tetraphenylporphyrin tetrasulfonate
MTT:	4,5-dimethyl thiazol-2-yl 2, 5-diphenyl tetrazolium bromide)
MV^+:	reduced methylviologen
MV^{2+}:	methylviologen

Na_2Pc:	disodum phthalocyanine
NADP:	nicotineamide adenine dinucleotide phosphate
NADPH:	nicotineamide adenine dinucleotide phosphate reduced form
Nd-YAG:	neodymium yttrium aluminum garnet
NEP:	nonaethylporphyrin
NHE:	normal hydrogen electrode
NiPc:	nikkel phthalocyanine
NMR:	nuclear magnetic resonance
OEP:	octaethylporphyrin
OMP:	octamethylporphyrin
OPO: o	ptical parametric oscillators
P_4Q_4Q':	quinone-linked porphyrin
PAHs:	polycyclic aromatic hydrocarbons
PBA:	pyrene butyric acid
$Pc(CH_2OC_{12}H_{25})_8$:	phthalocyanines substituted by long paraffinic chains
Pc:	phthalocyanine
PC_4VAC_4VB:	bis-viologen-linked porphyrin
$Pc-H_2$:	phthalocyanine
PcInCl:	indium phthalocyanine chloride
PcM:	metallophthalocyanine
PCnV:	viologen linked-porphyrin
PcV:	vanadium phthalocyanine
PDT:	Photodynamic Therapy
PdTCPP:	palladium tetrakis-(4-carboxyphenyl)porphyrin
PdTMPyP:	palladium tetrakis-(4-methylpyridyl)porphyrin
PdTPPS:	palladium tetraphenylporphyrin tetrasulfonate
PP:	protoporphyrin
P-PCnV:	para-viologen-linked porphyrin
PS I:	photosystem I
PS II:	photosystem II
PS:	polystyrene
PSC:	pressure-sensitive coating
PSP:	Pressure sensitive paint
PtOEP:	platinum octaethylporphyrin
p-TsOH:	p-toluenesulfonic acid
P-VA-VB:	bis-viologen linked porphyrin
PVC:	polyvinylchloride
PVS:	1,1'-dipropylviologen sulfonate
PVS⁻:	reduced 1,1'-dipropylviologen sulfonate
PyTP:	5-(4-pyridyl)-10,15,20-tritolylphenylporphyrin

RSH:	mercaptoethanol
Ru(bpy)$_3$$^{2+}$:	ruthenium tris(2,2'-bipyridine) dichloride
S:	singlet state
SCE:	saturated calomel electrode
SDS:	sodium dodecylsulfate
Si-OH:	silanol
Si-O-Si:	siloxane
τ_T:	lifetime of the excited triplet state
T:	triplet state
T$_1$:	lowest triplet state
T$_1$-T$_n$:	triplet-triplet
TAPP:	tetrakis-(aminophenyl)porphyrin
TCPP:	tetrakis-(4-carboxyphenyl)porphyrin
TLC:	thin layer chromatograpy
TMPyP:	tetrakis-(4-methylpyridyl)porphyrin
TPP:	meso-tetraphenylporphyrin
TPPC$_n$Br:	5-(4-bromoalkylpyridinium)-10,15,20-tritolylphenylporphyrin
TPPC$_n$V:	viologen-linked triphenylporphyrin
TPPS:	tetraphenylporphyrin tetrasulfonate
TPPSC$_n$V:	water-soluble viologen-linked anionic porphyrin
T-T:	triplet-triplet
TTC:	triphenyl-tetrazolium chloride
UV:	ultraviolet
VIS:	visible
Xe-lamp:	xenon lamp
Zn-TPPS$_3$$^{3-}$:	zinc tetraphenylporphyrin tetrasulfonate
ZnNEP:	zinc nonaethylporphyrin
ZnP(C$_4$V$_A$C$_4$V$_B$)$_4$:	water soluble bis-viologen linked zinc porphyrin
ZnP(C$_4$VC$_4$)$_4$:	butylviologen linked zinc porphyrin
ZnP(C$_n$V)$_4$:	water-soluble viologen-linked cationic porphyrin
ZnP(CO$_2$)$_3$CnBr:	water-soluble viologen free anioic zinc porphyrin
ZnP(CO$_2$)$_3$CnV:	water-soluble viologen linked anioic zinc porphyrin
ZnPc:	zinc phthalocyanine
ZnP-C$_{60}$:	fullerene-linked zinc porphyrin
ZnPC$_n$DQ:	bipyridinium salt linked porphyrin
ZnPC$_n$V:	viologen linked porphyrin
ZnPcSO$_4$:	zinc tetrasulfonatophthalocyanine
ZnPcTS:	zinc tetrasulfophthalocyanine
ZnPc-V:	viologen linked phthalocyanine
ZnP-H$_2$P-Im:	pyromellitimide-linked porphyrin

ZnP-Im-Q:	pyromellitimide-linked porphyrin with quinone
ZnSPc$_{mix}$:	Zinc phthalocyanines with different degrees of sulfonation
ZnTCPc:	Tetra-carboxylated zinc-phthalocyanine
ZnTMPyP:	zinc tetrakis-(4-methylpyridyl) porphyrin
ZnTPP:	zinc tetraphenylporphyrin
ZnTPPC:	zinc tetrakis-(4-carboxyphenyl) porphyrin
ZnTPP-PS:	zinc tetraphenylporphyrin-doped polystyrene membrane
ZnTPPS:	zinc tetraphenylporphyrin tetrasulfonate
ZnTPPS$^+$:	zinc tetraphenylporphyrin tetrasulfonate cation radical
^{3}ZnTPPS:	triplet state of zinc tetraphenylporphyrin tetrasulfonate
ZnTPyP:	zinc tetrakis (4-pyridyl) porphyrin
ZnTSPc:	Zinc tetrasulfophthalocyanine

Index

Milton Keynes UK
Ingram Content Group UK Ltd.
UKHW050452071024
449327UK00015B/344